Solar Energy for Man

B. J. BRINKWORTH

M.Sc.(Eng.), Ph.D., C.Eng.
M.I.Mech.E., A.F.R.Ae.S., M.Inst.P.
Senior Lecturer in Mechanical Engineering
University College, Cardiff

A Halsted Press Book

JOHN WILEY & SONS

NEW YORK · TORONTO

Acknowledgments for illustrations
Architects' Journal, 2, 3
Crown Copyright, 5
Daily Telegraph, 1
Middle East Navigation Aids Service
Bahrain, 7, 8
New Scientist, 4
USIS, 6, 9

To Harold and Alice Brinkworth

Library of Congress Cataloging in Publication Data

Brinkworth, Brian Joseph.
  Solar energy for man.

"A Halsted Press book."
Bibliography: p.
1. Solar energy. I. Title.
TJ810.B75 1973   621.47   73-549
ISBN 0-470-10425-2

CONTENTS

LIST OF ILLUSTRATIONS

between pp 76, 77

Eternal summer gilds them yet,
But all, except their sun, is set.
BYRON, *Don Juan*

PREFACE

There seem always to have been wide differences between nations in their supplies of food and of the material possessions which make for a comfortable standard of living. We, of the so-called Western nations, have been among the most prosperous largely because of our pursuit of science and our will to exploit what we have found. But a growing social conscience over the last century or so has demanded that we increasingly communicate our skills to less fortunate people. This has helped to slow down the widening gap of development, though it has not yet begun to close it; the very success of modern science has compounded the problem. A dramatic rise in the expectation of life in undeveloped countries has brought in its train the 'population explosion', giving the world each hour a further 8000 people to support. Many are born into such an atmosphere of hopeless misery that apathy swamps the will to change their state. Common humanity demands that we take this duty upon ourselves.

This is not another gloomy book warning of the difficulties. It deals rather with some possible ways of coping with them. One of the primary needs, in dealing with a growing population and lagging development, is for *power*. Power to produce things and to process them at rates far beyond the capacity of manual labour; power to synthesise materials and foodstuffs more rapidly than nature can; power to cultivate and irrigate the land; power for transport and communications; and so on. Most of the world's power has come hitherto from the energy stored in the fossil stocks of coal and oil, but the exhaustion of these stocks is already foreseeable and other sources must be considered. Nuclear plant can be expected to meet the large-scale demand in the long term. In this book, I am putting forward the case for obtaining energy directly from the primary source, the sun. This is more suitable for small, local demands in isolated places. Solar energy is plentiful in those countries most urgently needing power, though the difficulties in exploiting it are not inconsiderable. It is, of course, only one agency which might be used for power production, but there are many other ways in which the sun's energy can serve a human purpose—for heating, cooking, drying, distilling and so on. Solar energy is receiving increasing attention from engineers in many countries, and the principles of its exploitation deserve to be more widely known. Engineers have, in a sense, to live in the future, for their task is to anticipate human needs and to find ways of using scientific understanding in meeting these needs. The responsibility for our future well-being rests increasingly on them. It is impossible to believe that British engineers, who have already given so much to the world, will not have a part to play in meeting this new challenge.

This is not intended to be a text-book, though it contains many diagrams and not a few equations. It is not a comprehensive treatise and I have omitted references to original sources, though suggestions for further reading are given at the end. The initial incentive to write it came from the response of young people in school scientific societies to some lectures I have given on the subject of solar energy, under the auspices of the British Association for the Advancement of Science. It is addressed to sixth-formers, to undergraduate engineers and to the thoughtful reader generally. In trying to develop a style suitable for a diverse audience, I have aimed at putting forward nothing without explanation unless it is a statement of fact. To achieve this, I have tried to derive everything from first principles. I have, of course, had to assume a certain basic scientific knowledge on the part of the reader, so as to have a starting-point. But I have endeavoured to keep this, in general, within the compass of an average secondary education; only rarely shall we need to go beyond O-level work. The reader will find most of the results presented graphically. He may be asked to visualise some new concepts and to appreciate the results of some mathematical operations, but he need not be able to carry these out himself. It is my hope that no reader will be deterred by these modest requirements from gaining an insight into phenomena and processes which are certain to come into prominence during the next few decades.

Acknowledgements
Harold Heywood, then Principal of the Woolwich Polytechnic

and a notable worker in the field, first aroused my interest in this subject. David Rendel encouraged me to write about it and by gentle prompting has helped me to complete the text less slowly than I might otherwise have done. I may have been fortunate too, in the unwitting help of the Girl Guides Association, whose affairs have so monopolised my wife's evenings that I have been enabled to write most of this at home without trying her forbearance too far. Nevertheless, my debt to her is great in this, as in so many other ways.

Cardiff, 1969   B.J.B.

ENERGY AND THE HUMAN CONDITION

In a moment the ashes are made, but a forest is a long time growing.
LUCIUS ANNAEUS SENECA (*c.* 5 B.C.–A.D. 65)

OUR grandchildren will live in a world without oil. Coal will have become too precious for every day use as a fuel and will be reserved for the recovery of certain essential chemicals which it contains. The attendant changes in our way of life could be unpleasant, but the effects on the emerging nations will be worse. They may come to curse the reckless profligacy which will have destroyed the world's stock of these fuels before the majority of its people had reached a stage at which they could benefit from them.

So far, there has been no general concern about the problems that will arise from the disappearance of the traditional fuels. No doubt this is mainly because so few people are yet aware of the imminence of these problems. Even when they become self-evident, many people will merely make the usual comforting assumption that the specialists will find ways of solving them. In this case, the assumption is probably correct. It is the engineers' job to anticipate human material needs and some are already giving attention to methods of meeting this one. But there is little enough time to avoid adding to the world's list of shortages yet a further one—an energy shortage. The 'perpetual struggle for room and food', which Malthus foresaw in the early 19th century, could be joined by a new terror in a world so dependent upon the command of energy:

> *A thousand millions scrabbled the crust of the Earth. The wheels began to turn. In a hundred and fifty years there were two thousand millions. Stop all the wheels. In a hundred and fifty weeks there are once more only a thousand millions; a thousand thousand thousand men and women have starved to death.*

Thus Aldous Huxley, in the early 20th century. Even if we

manage to avoid this catastrophe, a mere shortage of energy would bring misery, injustice and acrimony enough.

Where can we look or deliverance? The emergence of nuclear fission as a power source has undoubtedly transformed the prospect. With its aid, a general crisis should be deferred long enough for the development of a fusion reactor to take place. If so, energy starvation will have been avoided, for an indefinitely long period of time. At present, this seems the only possible long-term solution. But it will not satisfy everybody. It will intensify the present trend towards power generation on a very large scale at central sites, with transmission to the user in the form of electricity. Only the wealthy nations will be able to afford the investment in research and capital equipment involved and the heavy cost of transmission. Poorer nations may not be able to match this investment and the development gap, which already divides the world, will continue to widen. For under-developed countries, the problem has a different aspect; their need is for modest amounts of power, in isolated and widely-separated places. Moreover, they need this *now*.

Those concerned with power supply and demand have been impressed by the relatively trifling duration of the period in which man can rely on the world's capital stock of the fossil fuels—coal and oil—compared with the known time-scale of his history. from time to time, studies have been made of the extent to which we could make use of energy *income*—solar radiation and the tides—and manifestations of these in the wind and river flows. This book is mostly about the use of solar radiation. We shall see that it is too diffuse and too variable to be a erious contender for power generation on a large scale though in principle it could provide us with all the power the world could conceivably need. For small, local requirements, on the other hand, it is undoubtedly attractive and the several methods for converting it discussed here are practicable now. We shall return frequently to its possible application in under-developed regions, where it seems that a modest effort could yield immediate and much-needed returns

In this first chapter, we shall look briefly at the scene in which this study is set and justify some of the statements already made, before narrowing the field to the possibilities of solar energy conversion.

1.1 *The human population*

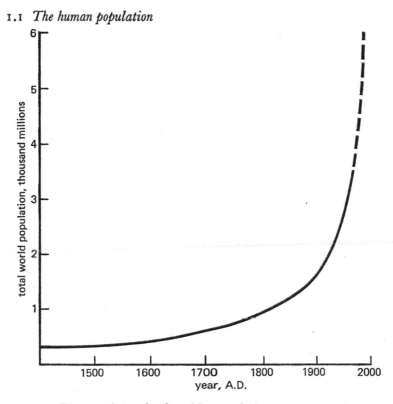

Fig. 1.1 Growth of world population, 1400–2000

It is generally recognised that in the twentieth century we are faced with an unprecedented and spectacular increase in our numbers, amply meriting the description 'population explosion'. Figure 1.1 shows some estimates of the total world population, with projections to the end of this century. Up to the 15th century, it is believed that the total population had remained in the vicinity of 300 millions for more than a thousand years. The change, beginning in that century, accelerates most strikingly until in our present century it takes on the aspect of an uncontrolled explosion. This is not just the result of a steady increase at 'compound interest'. Between 1400 and 1900, the *rate* of increase grew slowly from nearly zero to about ¾% per year. At the present time, however, we are increasing at a rate of nearly 2%

3

per year, adding in a single year a further number equivalent to the present population of the British Isles, or, in more mundane terms, about eight thousand extra persons every hour; *two people per second*. If this rate is maintained, the total world population will double again by the end of this century. Thus, in a mere 30 years time there will, in general, be two people wherever there is one person now. Just to maintain our present position we would have to double our rate of food production, extraction of raw materials and manufacture of consumable goods. Within those 30 years, we would have to build new houses, factories, schools and hospitals *equivalent to all those we possess now*.

An increase in numbers of at least this magnitude is now considered to be inescapable. How has this appalling prospect come about? Without any special restraint, it is found that a human population with a stable age distribution will produce between 40 and 60 births per annum per thousand people. (Demographers prefer this form of statistic to the alternative 4–6% of population per annum.) If the population is to remain constant, the number of deaths per annum would have to equal these figures. At an average death-rate of 40 per thousand, the average life-span of individuals would be 1000/40, or 25 years. For centuries, it appears that the birth-rate and average life-span remained not far from these values in most parts of the world and the growth of the population was slow. It was evident, however, from the occurrence in all periods of a fortunate few whose life far exceeded the average in duration, that man is potentially capable of living about 100 years. Much attention was paid, particularly in western Europe, to improvements in hygiene and medicine, which have gradually brought about a fall in the death-rate. In all the most highly developed countries today, the death-rate is down to about 12 per thousand, a value which, if maintained, would correspond to an average life-span of about 80 years. Moreover, the birth-rate in these countries has also fallen, to a value of abour 16 per thousand, so that the population continues to increase slowly.

The impact of Western medicine and principles of hygiene in less-developed parts of the world has brought about a remarkable reduction in the world's average death-rate in the last 30 years. Over this period an increase in life expectancy from 25 to 50 years has occurred in several countries, a change which took centuries to take place in Europe. This lengthening of life-span is mainly a consequence of a reduction in the mortality of infants. Where, formerly, only about half of those born lived to reach their teens, improvements in sanitation and the control of infectious diseases has raised this to more than 90%. This dramatic change has occurred in little more than one generation and social customs rarely alter so rapidly. The 'population explosion' is largely due to a delay in the fall of the birth-rate which must ultimately follow in these countries. The world-average birth rate is still about 35 per thousand, whilst the average death rate has fallen to about 17 per thousand. It is clear that until the birth and death rates are brought into approximate equality everywhere, the world population will continue to grow.

It is not known exactly how a society reacts to a rapid increase in size. The fall in birth-rate in western Europe has been slow and steady, though in post-war Japan it has been rapid and abrupt. One factor is the realisation that reduced mortality makes it unnecessary to have many children in order that the continuity of the family shall not be broken through death in infancy. We might suppose, also, that a multitude of social

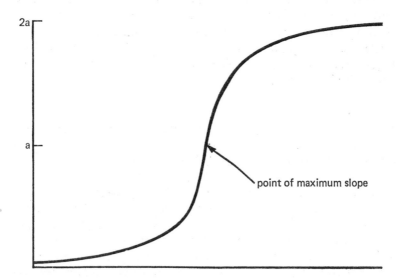

Fig. 1.2 The logistic of Verhulst.

changes make the having of a large family incompatible with having an acceptable standard of living. It has been found, too, that several kinds of animal population restrict their expansion when the population density becomes too high. The population size tends in these cases to follow the S-shaped curve, called the LOGISTIC by the Belgian mathematician VERHULST (1838), and illustrated in Figure 1.2. Human beings, with a conscious will, and access to numerous artificial methods of birth-control, need not necessarily follow a similar course, but it seems reasonable to expect that the pressures of the next decades will bring about a slackening in the rate of growth of approximately the same kind.

Since the logistic curve is anti-symmetrical, it will be seen that the ultimate value to which the population tends is twice the value at the point where the slope is steepest. We might assume for the moment that the world population will follow such a curve and that recognition of what is in store for us will result in the steepest point being reached within the next decade or two. We would then conclude from the data of Figure 1.1. that the population would not become stable until it has reached at least ten thousand millions. This is about three times the present level, but our assumptions are optimistic and it is, perhaps, the best that we could expect. Many fear that a steady level will not be reached in this way, but that a period of fluctuation of total population will ensue, with peak values much higher than our estimate, followed by rapid falls through shortage of food and other resources.

1.2 The human condition

We have seen how a rapid fall in mortality in under-developed countries, not yet followed by a fall in birth-rate, has caused their populations to increase rapidly. This has made even more acute the difficulties under which these peoples are obliged to live. Compared with these difficulties, the problems of life in developed societies, which demand so much of our anxiety and consume so much of our time and treasure, are almost grotesquely trivial.

Any measure of the plight of the under-privileged peoples is difficult to grasp. Few would disagree that the primary needs of any human being are for food, water, clothing and shelter, yet the majority of humans are inadequately provided with at least

6

one of these today. The food available to about *half* the world's population is lower, in quantity or in quality, than the minimum thought to be necessary for the maintenance of health and the capacity to work. Of these, perhaps one-sixth of the total population, some 500 *million* people, are dangerously under-nourished. The number whose food is inadequate in quality is even greater, and diseases caused by diet deficiencies are widespread. Hundreds of millions of children are afflicted with kwashiorkor, a protein-deficiency disease which enfeebles the mind as well as the body—yet could be eliminated at the cost of a few soya-beans per person per day.

Secure water supplies, uncontaminated foodstuffs and effective methods of waste disposal are available to less than a third of mankind. Aided by indifference through the lack of a tradition of hygiene, these deficiencies encourage the most widespread incidence of infection and infestation. Tens of millions still suffer from malaria; hundreds of millions from filariasis and bilharziasis; perhaps 500 millions from the virus-induced eye disease trachoma.

The tale of distress and misery extends to every field of human activity. It is thought that the domestic living conditions of more than a third of mankind are such as to constitute a health hazard. Many are entirely homeless, without work and without prospect of work. A fifth of the world's children cannot go to school, while some 800 million adults are unable to read.

Improvements in communications have begun to show to all the disparity between these conditions and those which man is capable of producing, as shown by the corresponding conditions in the developed nations. Although more than two-thirds of the world's population lives in Africa and Asia, the national incomes of countries in these areas amount to less than a sixth of the world total. North America, with about 7% of the world's population, has about 40% of the world's income. The purchasing power of an individual in America or western Europe is 20 to 50 times greater than that of his fellow in Africa or South America. The present pre-eminence of these few favoured nations is largely traceable to the vigour with which they have pursued the understanding and applications of science. Although the benefits of their knowledge tend ultimately to be communicated to less-favoured nations, the 'development gap' is widened

7

by many new scientific advances. Some of these improve the overall efficiency of the nation's economy and, by creating substitutes, reduce its dependence upon materials formerly imported. In this way, the less-developed nations, though becoming numerically stronger, tend to become economically weaker through a slackening of trade in such traditional exports as cotton and rubber.

It must not be thought, however, that nothing is being done to bridge the gap between developed and under-developed nations. International aid and the transfer of scientific, technical, agricultural and medical experts now take place on a very large scale. The activities of many of the agencies of the United Nations Organisation, advised by many of the world's leading authorities, are directed entirely towards reducing the inequalities between nations. All that Western civilization has learned in overcoming hunger, disease and hardship among its own peoples can be brought to bear on these burdens wherever they occur. Moreover, the signs of a growing social conscience in the favoured nations promise that the scope of these applications will continue to widen. Indeed this must happen, even if prompted by nothing more worthy than self-interest. For if it does not, there can only ensue an era of desperate violence and a recession of culture and ethics.

Nevertheless, it is far from certain that economic development everywhere must follow a pattern similar to any of those which have been successful in Europe, the U.S.A. and the U.S.S.R. The problems of food and water supply, housing, clothing, public health and education, though interconnected, vary much in degree from place to place. So also do the needs and aspirations of the people.

We cannot enter into all these matters here, but will concentrate on but one aspect of one of the associated factors, which many have come to feel to be a central one: the provision of *power*. The increase in living standards, through which Europe first began to draw away from the rest of the world, stemmed directly from the Industrial Revolution which began in Britain about 150 years ago. Because of this example, mechanisation and other associated technological developments are being seen as a key to a general improvement in living standards. If this is so, the solutions of many of the problems we have considered are

8

dependent upon the provision of power on a quite unprecedented scale.

1.3. The demand for power

The utilisation of power by man in the past has followed a similar trend to the growth of population. Until about the 16th century, heating was provided universally by the burning of wood, agricultural wastes and animal dung, and work was performed either by man himself or by animals. The beginning of the use of water power and windmills is of great antiquity, but these also used replaceable sources and made no inroads upon the earth's capital resources. Sea-coal from occasional outcropping strata was used in Britain as early as the 12th century, but the production of coal by mining did not reach ten million metric tons per year (1 m.ton = 1000 kg) until the end of the 18th century. At this time, the industrialisation of society began in Britain and the demand for power soon passed for ever beyond the level which could be provided by replaceable sources alone. From SAVERY's steam-pump of 1698 was developed NEWCOMEN's engine (1712), followed by WATT's engine with crankshaft (1781) and TREVITHICK's locomotive (1802). With the invention of the electrical generator by PIXII (1832) and GRAMME's electric motor (1873), following FARADAY's discovery of the laws of electromagnetism, the way was opened for a massive growth in the use of power. By 1900 the amount of coal mined in Britain alone had exceeded 200 million metric tons per year. Oil began to be extracted in quantity in Romania (1857) and soon afterwards in America (1859). As well as replacing coal as a fuel for chemical processes, oil has been much in demand as a source of refined fuels for internal combustion engines, whose origins in the field of transportation also belong to the late 19th and early 20th centuries (OTTO's petrol engine, 1875; BENZ's motor car, 1890; WRIGHT's aeroplane, 1903).

To view the rise in demand for power, that is, the rate of utilisation of energy, in a suitable perspective, we need to make an estimate of the extent to which man has learned to augment his own meagre capacity for work. We may start from the statistics for the total world production of coal and oil. Nearly all of this will have been burnt with oxygen from the air and its chemical energy employed either for heating homes, public buildings and

factories or for operating engines in which it is ultimately converted into work. The efficiency of these processes has varied
with time as a result of technological improvements and social
changes. Where fuels were used for heating, the efficiency of
utilisation has risen quickly in the last few decades, but, as we
shall see in Chapter 5, there are limits of a fundamental kind on
the efficiency of engines. We need not try to be too precise, since
our purpose is only to obtain a rough impression of the demand
for energy, so we will probably be justified in assuming that the
average efficiency with which all coal and oil has been used up
to now is no more than 20%. If so, the *demand* for energy has been
about $\frac{1}{5}$ of that potentially recoverable by burning these fuels.
To this must be added an allowance for processes in which
wood, peat and other vegetable matter are used. These are still
the major sources of energy in many parts of the world and have
been estimated to represent about 15% of the total fuel requirement today. In the past, their relative importance will have been
greater. We will suppose that the proportion of the world's
demands met by these fuels was 50% in 1900 and that their

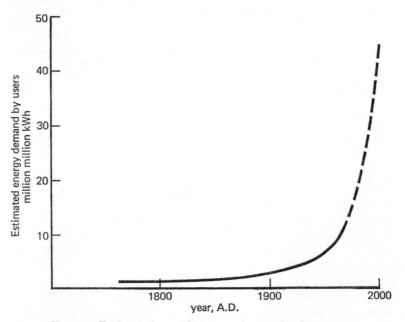

Fig. 1.3 Estimated world energy demand, 1800–2000

total energy contribution had remained at about the same level before that time. The contribution from other sources, such as hydroelectric and nuclear power plant is still quite small, probably not more than 2% of the total demand, though we need not make much allowance for efficiency of usage because this is high for all processes using electricity. Although the number of animals used for load-carrying and agricultural work must run into hundreds of millions, the total quantity of work performed by them is quite negligible on the scale we are considering here, for the average output of such animals is probably no more than a few kilowatt-hours per day.

Given these premises, the total world demand for energy has taken the form shown in Figure 1.3. The unit of energy employed in this figure is the kilowatt-hour (kWh), the familiar 'unit' for which we are charged in our domestic use of electricity. It must be emphasized that this figure represents a crude estimate of the actual work performed by man-made machines, with the energy used in chemical processing and for heating. We may call this the ENERGY DEMAND. It must not be confused with the energy supply; to meet the demand, the supply must be much greater, because of the inefficiency of the processes by which the demand is realised.

The curve for energy demand steepens more rapidly than that for the world population, Figure 1.1. This is readily understandable. As a desire for a higher standard of living grows in a previously undeveloped society, so does its energy demand increase, per head of population. The pattern which has emerged in Western nations is a movement from life at mere subsistence level, in which most of the population is engaged in agriculture, to a progressive mechanisation and urbanisation in which the majority of people, as well as the generation of power and the manufacture of goods, are centred in towns and cities. The material advantages which accompany this process are plain to see, and although it should not be assumed that all peoples desire these advantages, it has been said that the state of development of a nation can be judged from the proportion of its people which lives in urban areas. In Britain today, about 6% of the population is engaged in agriculture, compared with nearly 90% in some Far Eastern countries.

Another telling indicator of material development is the

usage of energy per head of population. The amount of work that an adult human being can perform by his own effort alone is very small. For the maintenance of fitness, he must have a balanced diet whose energy content is about 2 500 kilocalories per day, corresponding, in the units we are using, to about 1 000 kWh per year. Of this, he cannot furnish more than about 5% or 50 kWh per year, in useful work. For a population as a whole, the average expenditure of energy in work could not be more than about half of this. With the assistance of animals, one man can command much more energy and with the development of energy-converting machines, more again. The average energy demand per person for the world as a whole has already exceeded 3 *thousand* kWh per year. There are, however, very wide differences in the average energy demand from nation to nation. In the United States and parts of Western Europe, it has, on our definition, reached about 18 thousand kWh per year, whereas in India it is still only a few hundred kWh per person per year.

It will be the aim of developing countries to raise their standards of living until the present glaring disparities have been eliminated. How much further the development of richer nations will continue whilst this happens is entirely conjectural. We might at first suppose, so as not to overestimate too grossly, that a steady state could be reached in which the average energy demand per head of population for the world as a whole had risen to about the same level as in the most highly-developed nations today. This would allow for some further development by the richer nations if it is also supposed that not all peoples would desire the heavily-industrialised environment to which these nations are now committed. However, there are many additional factors to be taken into account. Advances in presently under-developed countries will demand capital construction on a large scale in a short space of time. The requirements for raw materials to support this expansion would soon exhaust the world's mineral resources if these were reckoned on the basis of present methods of recovery. In these terms, the recoverable reserves of certain metals are already measurable only in decades of further extraction at present rates. An increased demand will mean extraction from ores in which these minerals are present in much lower concentrations than are considered to be workable today. For this, more energy will be needed to recover a given quantity of mineral in the future than in the past. Conservation of certain metals by repeated re-use, and their replacement by plastics and other substitutes will not much reduce the demand for energy in the provision of the materials of construction.

Water, too, will no longer be available cheaply. There are already areas where every drop of rain that falls is committed in advance for some human purpose. It is certain that the demand for water in the future will necessitate the desalination of sea-water on a vast scale, for which the energy requirement will be a new factor in the total demand. For the cheapest method currently available, this is about 100 kWh per metric ton of water, and thermodynamic limitations will prevent this from being brought below about 2 kWh per ton. Further, it is impossible to foresee the feeding of the probable population of the world without a greatly increased use of fertilizers, perhaps at a rate of a hundred times the present levels. Energy is needed in large quantities in the extraction and manufacture of these. The most crucial process is the initial 'fixing' of nitrogen in the preparation of ammonia, NH_3, from which many other derivatives are obtained. If the nitrogen is recovered from the air and the hydrogen obtained by electrolysis of water, about 7 kWh of energy is required to prepare the equivalent of 1 kg of fixed nitrogen. Already the world production of fixed nitrogen in fertilizers exceeds 20 million metric tons per year.

These factors, and many others, make any estimates of the ultimate world demand for energy speculative in the extreme. However, it is hard to feel that this ultimate demand could be less than about 20 000 kWh per person per year. If we combine this with our hopeful estimate of the total population in this steady state, we arrive at an ultimate energy demand of about 200 million million kWh per year; about 16 times the present level. We note that this increase arises from a postulated increase in population by a factor of about 3 and an increase in average energy demand per head by a factor of more than 5. Based as they are upon crude assumptions, these figures cannot be relied upon to give us more than a general impression of possible trends. But it must be hoped that they do not seriously underestimate the demand, for we cannot expect to be allowed more than a century in which to meet it.

1.4 Meeting the demand

The provision of the world's energy requirements cannot properly be considered separately from the attendant problems of providing man with food and water, supplying his industries with raw materials and disposing of his wastes. All these have reached, or are about to reach, a critical state. We cannot examine all the implications here, for the book has another object. It will suffice to note that an energy requirement such as we have just deduced cannot be provided by our stock of fossil fuels.

Men have been prospecting for coal for over a century and it is certain that the reserves are mostly known. Subject to the condition that it should be recoverable by techniques not greatly different from those used today, the remaining stock is generally reckoned to be in the vicinity of $2\frac{1}{2}$ million million metric tons. If utilised with an average efficiency of 30%, this would meet, in total, a demand of about 6000 million million kWh. At the level of demand which can already be foreseen for the end of this century, this represents less than 150 years' supply; at the ultimate rate of demand forecast in the last section, it represents a mere 25 years' supply. Of course, we shall not then rely only upon coal, but these figures show clearly how near we are to exhausting our capital.

The contribution from oil and natural gas is more difficult to estimate. The geological features indicating their presence are not so clearly defined as for coal, and estimates have to be based on predictions of the probable future rate of discovery of new reserves on the basis of past experience. Reserves known today would meet an energy demand of about 600 million million kWh, and probable future discoveries are not generally expected to yield more than about three times this value. With so little available, we must expect the supply of oil and gas to be exhausted quite soon. They are more convenient fuels than coal and have increasingly taken its place in power generation and heating plant. Already we consume in a single decade more oil than all that which has been produced previously. If no more than half the estimated world energy demand shown in Figure 1.3. were provided by oil, most of the known reserves would have been used before the end of this century.

Faced with the very brief duration of our possible dependence upon fossil fuels, it is hard to be unimpressed by the timeliness of the discovery and exploitation of fission energy. Since the first achievement of a fission chain reaction by FERMI and his colleagues in 1942, the development of nuclear reactors has proceeded steadily, notably in the United States and Britain. The fuel for these reactors has mostly been 'enriched' uranium metal, in which pre-treatment has raised the concentration of the active isotope U–235 above the naturally-occurring level of 0·7%. However, it has been found that the more abundant isotope U–238 and a commonly-occurring isotope of thorium, Th–232, can be converted into the fissionable isotopes Pu–239 and U–233 respectively. This process takes place in a 'breeder reactor' in which the conversion to fissionable material is brought about by energy from the fission of U–235. The quantity of new fissionable fuel produced is somewhat, though not greatly, more than that used to power the reactor. A more promising prospect is the development of a 'fast' reactor in which the common fissionable isotopes can be used directly. At present, uranium is recovered only from rocks in which the concentration is greater than about 1 part in 1000, but in an energy shortage it will become economical to recover it from rocks having much lower concentration than this. Thus it is difficult to estimate the probable reserves of potentially fissionable material, but they must certainly be reckoned in tens, if not hundreds, of millions of metric tons. If all uranium and thorium can be made to undergo fission, one metric ton of either would be expected to lead to the production of at least 100 million kWh of electricity. Since the generation of electricity in coal-fired power stations yields only about 2500 kWh per ton, the gain is quite striking. It can be visualised another way. The abundant rock granite, which has a world-wide distribution, contains on the average about 20 parts per million of uranium and thorium. Should it become possible to recover and use this, even at the rate given above, every ton of raw granite could give as much electricity as 8 tons of coal.

We need not speculate far beyond this point. It is evident that nuclear power is potentially capable of deferring world energy starvation for a very long time and that it is being developed at the right moment to be ready to take over the bulk of the energy supply as coal, oil and gas become exhausted. Beyond this lies the further prospect of using fusion reactions as power sources, employing the heavy isotopes of hydrogen, recovered from sea water. Indeed, the seas may ultimately yield mankind all that we might need of food, water, fuel and minerals, so that in the long term there seems to be no fundamental reason for pessimism about the earth's capacity to support us all.

1.5 Energy income

So far we have considered only the capital resources of the earth. We cannot live on capital for long and we must ultimately learn how to live within our income. As we have seen, all the world's fossil fuels may be used up over a period of a couple of centuries; a mere instant in the history of the planet. These fuels are considered to be the residue of plant life from some period when conditions allowed this to grow in great profusion everywhere. Only in the tropical rain forests is material being laid down at remotely comparable rates today, but their eventual contribution to our fuel capital is entirely negligible.

The energy required to form this material comes, of course, from the sun. Light-induced reactions in plants, of the kind called PHOTOSYNTHETIC, which we shall examine again in Chapter 9, enable them to manufacture organic (carbon-bearing) compounds. The raw materials for these reactions are water and carbon dioxide. All life is driven by the energy recovered when organic chemicals are broken down again and these basic substances reconstituted. In these universal processes, energy is transferred on a prodigious scale. It is estimated that plant life on land alone utilises more than a hundred million million kWh of energy per year; no reliable methods are available for estimating the corresponding figure for life in the seas, though this must be the of same order of magnitude. Thus, the total rate of capture of solar energy by living organisms is certainly many times greater than the energy demand of man at the present time. It is chastening to note, however, that it is *not* much greater than the minimum value of our probable ultimate demand, as estimated in Section 1.3. In any case, the fraction of the solar energy captured by photosynthesis which could be recovered by man is very small. It has been said that, long before the end of the 19th century, our energy demand had passed the point at which it could be met by burning timber, even if forestry were developed to the highest conceivable degree.

16

Solar energy enters the earth's economy by another route. By warming the air, the sea and the land, it provides the energy to maintain the circulation of the atmosphere and the oceans and to evaporate water, which is to return as rain. After falling on high places, rain can be made to operate hydroelectric plant on its return to sea level. In Switzerland and some Scandinavian countries, more than half the energy demand is met in this way. Though the overall contribution of hydroelectric power to the total world production is quite small, several under-developed areas—notably in Africa and South America—are well endowed with potential hydroelectric sources. Even so, it has been shown that, if developed to the greatest feasible extent, the world's capacity for producing water power would not exceed our present total demand. It seems that we cannot expect a major contribution from wind power either. Extensive studies have been made of potential sites for advanced windmill generating systems and some quite large ones, with capacities in excess of 1000 kW, have been built. These are unquestionably very valuable devices for modest power generation in remote areas and their use is certain to grow very greatly. Even if they were used on the widest conceivable scale, however, the potential power supply from wind driven generators is not believed to be greater than that from hydroelectric plant.

Generating systems sited at river estuaries may use partly the energy of the river and partly that of the tides in the sea. Tidal power plants have been much studied and some very large ones have been built, notably the Rance barrage on the north-western Atlantic coast of France, with an annual generating capacity of about five hundred million kWh. The tides are resisted by friction on the beds of estuaries and beaches, which has the effect of slowing down the rotation of the earth. It is known that the earth's day is lengthening at a rate of about one second in 120000 years, so that the energy transfer involved can be calculated. It is far less than even our current energy demand, and only a minute fraction of this could be intercepted, for the rise and fall of the tides is not everywhere large enough to make recovery economically attractive.

There is room for development of all of these methods of utilising the world's energy income. Each has its own special attractions and disadvantages. In this book, we shall restrict our study to the direct use of the sun's radiation. Though there are many aspects to examine before we can see the extent to which this could lead to real economic advancement, we can quickly get an idea of the possibilities. At the top of the atmosphere, the intensity of solar radiation, that is, the rate at which it carries energy per unit area of exposed surface, is about 1·3 kW per square metre. This is about the equivalent of the average electric fire for an area equivalent to the average table-top. For reasons which we shall discuss later, much of this fails to reach the earth's surface, but there are many regions in which the total energy falling on a horizontal surface exceeds 2000 kWh/m² per year—equivalent to an average intensity of 0·6 kW/m² for 9 hours a day. A region 80 km square would thus receive as much energy as all mankind uses at the present time, and one 300 km square would supply even the inflated demands of the future. To get these figures into perspective, we might note that the desert region of Western Australia is about 1500 km square.

Striking though these figures are, we must recognise that solar energy is, nevertheless, very diffuse by normal engineering standards. Even in a domestic kettle, for instance, the rate at which energy is transferred across the heating surface is several *hundred* kW per square metre. The low intensity of solar energy is one of the biggest obstacles to economic recovery. Yet there can be no doubt that, had we not learned how to obtain nuclear energy, sooner or later we would have been forced to recover solar energy for nearly all our purposes. For it is the only re-current source large enough to meet man's demands.

1.6 *Patterns of development*
Though we shall not now need to develop solar energy recovery on a large scale, it is very attractive for more modest local requirements. It occurs wherever it is needed, in a certain and inexhaustible supply. We do not have to think of receiving areas many kilometres square. All the hot-water needs of a hospital can be provided by the solar energy which falls on its roofs. Acting as an energy collector, the roof of a house can provide power for a loom or a lathe; the roof of a barn can power a pump, able to irrigate ten hectares (25 acres) of land. A solar still the size of a duck-pond can yield enough drinking water for a whole village. A school's radio and television receivers can be powered by a

device the size of a blackboard—the possibilities of such an on-the-spot supply are endless.

But what does this mean in terms of the economic development of under-privileged regions ? Its greatest potential, seen in the long term, is probably the opportunity it affords to delay the drift of population from rural to urban areas. Thereby, it opens up the prospect of a different pattern of development from that of wholesale urbanisation, the evils of which are adding another appalling burden to the developing nations.

The first step in technological development is to free people from an existence at mere subsistence level, in which each family is occupied solely in producing its own food. In Europe, this happened gradually, but can be traced to two concurrent processes. Firstly, improved methods of farming have vastly raised the productivity of an individual working in agriculture, so that a progressively larger fraction of the yield became available for consumption elsewhere. Nowadays one man can produce enough food for tens of others. Secondly, the people thus released have become concentrated in towns and cities where they can be efficiently employed in manufacturing industry. The products of this industry contribute to a general rise in living standards and many of them are sold overseas to people who can use them, but cannot make them economically themselves. The worst excesses of labour exploitation which at first accompanied these processes have given way to a more equitable distribution of the wealth created. Gradually a society has grown up in which a large fraction of the people can be allowed to engage in socially desirable, but economically non-productive occupations. These have provided the developed nations with those things which fulfil their conception of a civilised life—good general medical care, universal education, facilities for leisure, and so on. It is noteworthy that these amenities are a feature of life in all developed countries, whatever their predominant ideology, and that in general, they seem to figure in the aspirations of any society.

The processes described have brought about a relative depopulation of rural areas. In England, for instance, the number of workers employed on the land has fallen by a factor of ten in the last hundred years. The vigour with which industrialisation was embarked upon and the haste with which workers were gathered into the towns in the 19th century led to much misery and squalor. The remnants of this have yet to be eliminated entirely. But the scale upon which it is being repeated in the emerging nations is already much greater and the consequences pitiable in the extreme. Shanty towns, growing up on the outskirts of cities, sometimes hold a third of the city's population. Dwellings, built of waste material like packing cases, flattened tins and paper, are put up wherever space can be found, with no roads, water supplies or sewers. These one-roomed hovels are badly crowded, with one person to 5 m² of floor space or less. The occupants are transient, illiterate, enfeebled and under-employed. Yet the population grows year by year, beyond all hope of re-housing. Already, more than a quarter of the world's population lives in or near a sizeable urban area. On present trends, this would be expected to reach a *half* by the year 2000.

People press in like this because even these terrible conditions are better than those in the country. The harshness of traditional rules of land tenure in some countries has, in itself, been enough to urge any self-respecting man to move out. The city is a magnet, with its vague attraction of somewhat more work, nearness to amenity, contact with a larger world, a feeling that things are more likely to happen there than elsewhere. It is true that such resources as are available are invariably expended mainly in the urban areas. Without deliberate action, there is no reason to suppose that the attraction of the city will diminish.

Rural depopulation is proceeding in some places at a rate much exceeding that which might be justified by the increase in agricultural productivity. As a result, hardly any surplus food is being produced, beyond that required by the grower's family and neighbours. The break-up of family life occasioned by the drift to towns further diminishes the stability of the rural community. These and a multitude of other reasons demand that the movement of people out of rural areas must be arrested. For this to happen, work and amenity must be provided in the village. Lack of communications is a serious obstacle to this. Most of the minor centres of population are quite unconnected with each other. Road and rail systems, electricity grids and oil pipelines cannot be provided economically in large countries with low population densities.

The problems of development of remote areas are many and interconnected. We cannot pursue these in any detail here, but it suffices to say that one of the keys to a solution is the provision of power on the spot. Generation of electric power at large central stations where this can be done economically is unhelpful, for the cost of transmission over large distances is very heavy. The 'doubling distance', that is, the distance from the source at which the cost to the user is doubled, is only about 500 km for electricity. Generation by burning fossil fuels on the spot is hampered by their scarcity in many underdeveloped countries and again by the cost of transporting fuel. The doubling distance for coal is also merely a few hundred kilometres.

This is why wind and solar power are so interesting at this stage. They are available in useful quantities nearly everywhere. We shall see that the most economical size for solar plants is small rather than large and that they can provide on the spot the sort of power needed for local enterprises such as weaving, timber sawing, paper milling, food preserving, light manufacturing, vehicle repairing, irrigation, fresh water supply, drainage and so on. If villages can be re-vitalised from within, by the establishment of minor industries and amenities like these, they will gradually link into larger regions with greater opportunities for economic advance. This is now seen to be a far better solution than to attempt to provide large urban and industrial centres on the Western pattern in a short space of time.

We have said that many places receive solar energy at a rate exceeding 2 000 kWh per year for each square metre of surface. The rest of this book is given to the examination of methods by which this energy can be recovered and utilised. It is obvious that these methods have some disadvantages, or more of them would already be in universal use. In considering these disadvantages, we shall try to determine which methods are likely to be the most fruitful. Perhaps it can be anticipated now that the answers, as in most matters of technical economics, are not clear-cut. But the object of the book will be achieved if more people are made aware of the possibilities of these methods and of the crucial contribution to economic development of which they are potentially capable.

2

THE SUN AND THE EARTH

The great world of light, that lies behind all human destinies.

HENRY WADSWORTH LONGFELLOW (1807–1882)

IT will be appropriate to begin our study of the use of solar energy and its possible benefits with an examination of the raw material involved. This is the radiation which streams out ceaselessly from the sun into space and of which a minute fraction is intercepted by the earth. It is the fuel for all earthly processes, animate and inanimate, (with the exception of trifling contributions from nuclear reactions and gravitational interactions leading to the tides). In this chapter, we will examine the origin of this radiation, its nature and its interaction with the earth and its atmosphere. We shall determine how, for any point on the earth's surface, the intensity of solar radiation varies with the time of day and season of the year.

In later chapters, we shall relate this basic information to possible methods of using solar energy. Our purpose will be to understand the principles upon which these methods work and to be able to predict, with quite simple analyses, the efficiencies of each method of energy conversion. In doing this, we shall have to move across the boundaries of several disciplines and to use some concepts which are not yet generally familiar, in the fields of thermodynamics, quantum mechanics, electrochemistry and so on. The reader should not be alarmed at this. At every stage it will be our purpose to examine these concepts and to relate them to more familiar things. A simple explanation will be sought for every phenomenon described.

In general, this examination of concepts and their relation to fundamental principles will be undertaken whenever the need arises in the unfolding of the overall study. Between this chapter and the next but one, however, a short chapter is inserted which is devoted entirely to a consideration of fundamentals before the study is resumed. It is not obvious where such an insertion should be made. Already we have used a number of concepts which need to be thoroughly examined, and in this chapter we shall introduce more. The reader who finds the story unfolding without giving too much difficulty might feel ready to skip the chapter which follows this one, so that continuity will not be broken.

2.1 *The origin of solar energy*

Although it is unique and vitally important to us because of its nearness, the sun is an unremarkable member of what is called the MAIN SEQUENCE of stars. It belongs to the class of dwarf yellow stars, of spectral type dG, whose members are more numerous than those of any other class. These stars, like the others, are currently supposed to have been formed from a tenuous cloud of gas having an earlier existence. In such a cloud, random movements of material would give rise to isolated regions of density greater than the average. Matter would then begin to concentrate round these regions under the action of gravity. This process is called COLD ACCRETION. Any such assembly of matter will draw in upon itself, tending to states of progressively increasing density under its own gravitation. Eventually, the condensation will cease when it is opposed by forces between the particles which come into play when they come close together. We shall have more to say about these forces later.

The gravitational forces between the particles in such a cloud draw them together as if they had been connected by stretched springs. It does not matter for the moment that, to represent gravity, these imaginary springs would have the unusual characteristic of applying a force which gets weaker the further it is stretched. The important point is that as the particles are drawn together, they will have to move faster. The work done by the springs as they shorten (representing the energy change of the gravitational field) appears as energy of motion of the particles. Each particle is, of course, interacting with a multitude of others. The general state of agitation of the particles in the cloud rises as a result—we say that the temperature has risen. At some point, the temperature rises enough (the particles can approach each other closely enough) for certain nuclear reactions to occur. The ensuing events depend upon the nature and amount of material present. In the case of the class of stars to which the sun belongs, they enter into a long period of relative stability, in which they are contracting very slowly, the tendency to collapse being nearly balanced by the pressure of outgoing radiation.

Near the centre of the sun, the temperature is estimated to be about ten million degrees C and the corresponding motion of the matter to be so violent that it can no longer retain the ordered structure of atoms and molecules familiar to us. It becomes a PLASMA, in which the nuclei of atoms move about independently of their associated electrons. The nuclei frequently collide and interact, leading to fusion reactions of the kind that occur momentarily in thermonuclear explosions. It is possible, by observing the cooler material near the visible surface of the sun, to infer what nuclear reactions are in progress deeper inside.

The temperature at the visible surface is about 5500°C, still high enough for atoms to be in a very excited state, but low enough for them to take up occasionally the atomic forms known on earth. These forms are identifiable from the ways in which they interact with light. We must look at this later. For the moment, it suffices to say that, from the absorption and emission bands in the solar spectrum, characteristic of the atoms there, about two-thirds of the elements found on earth have been shown to be present in the sun. But by far the most abundant is the lightest, hydrogen. This constitutes some 80% of the sun's matter, nearly all the remainder consisting of the next lightest element, helium. It is, therefore, commonly agreed that the principal source of the sun's radiant energy is the fusion of hydrogen nuclei, which leads to the formation of helium. The nuclei of hydrogen are single particles called PROTONS, each carrying a positive electric charge. Similarly-charged particles normally repel each other, but if the temperature is high enough, they can have a sufficiently vigorous motion to allow them to approach very closely, to the point at which a short-range attractive force takes over and they can undergo fusion. The overall result of the chain of events which follows is the transformation of each four hydrogen nuclei into one helium nucleus, with the liberation of two particles called NEUTRINOS and some gamma radiation. The neutrinos are very unreactive particles, which are able to escape out of the sun altogether, and take no further part in events, though the gamma radiation does not escape unchanged, as we shall see. For the process as a whole, there is a net loss of mass, amounting to about $\frac{3}{4}\%$ of that of the matter taking part. As a result, the mass of the sun is decreasing at the astonishing rate of about 4 million tons per second, but such is its size that it is nevertheless expected to be able to remain in its present state for some thousands of millions of years to come.

2.2 *The solar radiation*

The nuclear reactions responsible for the sun's radiations occur in a central core, perhaps occupying only 3% of its volume. The surrounding material, some 500 000 km thick, profoundly modifies the outgoing radiation, which begins as gamma radiation. All electro-magnetic radiation is similar in nature, differing only in wavelength. Gamma radiation has the shortest wavelength known—of the order of a hundred-millionth of a millimetre. Now the ways in which radiation interacts with matter are most easily understood if we suppose radiation to exist in discrete particles, known as PHOTONS, each having a fixed amount, or QUANTUM, of energy. We shall discuss this idea more thoroughly in Chapter 4. We need only say now that, in this model, the energy of a photon is found to be inversely proportional to the wavelength of the radiation represented by it. Then gamma radiation photons are extremely energetic because of the short wavelength. As they stream outward from the centre of the sun, they collide with nuclei and electrons or are scattered in near-collisions. At each event, some of the energy is shared out with the particles struck and the remaining photons of radiation, having less energy than before, have a longer associated wavelength. The gamma radiation is soon softened into X-rays, with wavelengths in the region of a millionth of a millimetre. Nearer the surface of the sun, where the temperature has fallen sufficiently for whole atoms to exist, there is a further mechanism for softening the radiation, in which the colliding photon moves an electron from an inner to an outer orbit. The electron, returning from this excited condition to its normal one, gives up the energy again in radiation of a characteristic frequency. Finally collisions give whole atoms extra kinetic energy, which is shared out in the turbulent motions of the gas at the surface. The radiation leaving the sun and passing out into

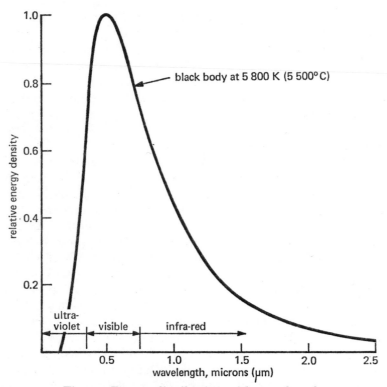

Fig. 2.1 Energy distribution with wavelength

space is now distributed throughout a wide range of wavelengths with most of the energy between a ten-thousandth of a milli-metre and a hundredth of a millimetre. The energy distribution is fairly close to that of the classical 'black body' at a temperature of about 5500°C, as illustrated in Fig. 2.1. We shall have to dis-cuss the black body again, but should note the shape of the curve here. The ordinate of the figure is the ENERGY DENSITY, defined in such a way that the area beneath the curve between any two wavelengths is proportional to the energy carried by the radiation in that part of the spectrum. The unit of wave-length used in the figure is the micrometre or micron, equal to a millionth of a metre (i.e., a thousandth of a millimetre) and written μm. It can be seen that about a half of the sun's energy is radiated with wavelengths between 0·35 and 0·75 micrometres, that is, in the VISIBLE BAND. Thus, we might note in passing

26

that man has evolved—or was created—so as to be able to see by the kind of radiation in which the sun's output is most abundant. There is very little radiation in the ULTRA-VIOLET band, below 0·35 micrometres wavelength, but more in the INFRA-RED, above the visible band, contributing to our warmth, though we cannot see by it.

With the aid of a piece of smoked glass or welders' goggles, or by projection of an image of the sun through a pin-hole onto a plain surface, we can readily see that the sun's visible outer envelope is not uniform. The dark areas—the SUNSPOTS—are the most obvious features, but closer study at higher magnification reveals a seething, granulated surface dotted with lighter (hotter) regions—the FACULAE—erupting into gigantic spouts —the PROMINENCES—hundreds of thousands of kilometres in extent. Such a state of turmoil exists at the visible surface that the relative constancy of the outgoing radiation is rather surprising. Both the total intensity and the distribution by wavelength seem to vary very little, such effects as are noticed being greatest in the ultra-violet region. The upheaval in the sun's surface is shown most strongly in yet another phenomenon, the SOLAR WIND. This is an irregular stream of matter, consisting mostly of protons, which has acquired enough energy to leave the sun entirely, passing out through the solar system with velocities of hundreds of kilometres per second. The interaction of this stream with the earth's magnetic field causes the auroral displays and interference with radio communications.

2.3 Interaction with the earth's atmosphere

About eight minutes after leaving the sun's surface, its radiation, travelling at 300000 km per second, has reached out to the orbit of the earth. It is now passing through a surface 150 million km (93 million miles) in radius, but so great is the sun's power output that the intensity per unit area, even at this distance, is about 1300 watts per square metre. Though the earth intercepts but a minute fraction of the sun's energy, that falling on our upper atmosphere in a given period is equivalent to tens of thousands of times the present energy requirements of the entire world in that period. Not all of it reaches the earth's surface, however, as a result of a number of complex interactions in the atmosphere, illustrated in Fig. 2.2.

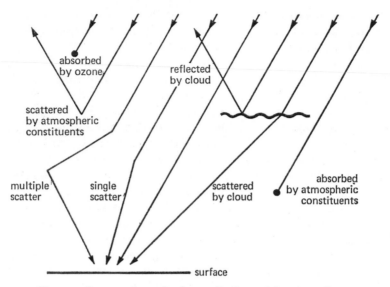

Fig. 2.2 Interaction of solar radiation with atmosphere

At the highest levels, above some 25 km from the surface, a process occurs which removes virtually all the ultra-violet radiation. Ordinary molecular oxygen, O_2, is first dissociated into atomic oxygen, O. The energy required to break the molecular bond is such that only photons with wavelengths less than about $0.18\mu m$ are effective and all this radiation is removed in the process. Some of the oxygen atoms recombine into molecules again, but most react with other O_2 molecules to form molecules of ozone, O_3. This is another strong absorber of radiation, but not so strongly bound together as O_2, so that photons with wavelengths less than about $0.32 \mu m$ can dissociate it into O and O_2 again. A steady state now exists in which dissociation and recombination of O, O_2 and O_3 occurs continuously, removing virtually all the ultra-violet radiation and transforming it into photons of lower energy. This is a fortunate circumstance for us, since ultra-violet radiation burns the skin, damages the eyes and can even be lethal in moderate doses.

Photons with wavelengths in the visible and infra-red bands interact with gas molecules and dust particles in the air without being able to produce bond fracture. Instead they are scattered, more-or-less uniformly in all directions, so that some of the radiation is re-directed away from the earth and back into space again. This type of scattering affects the short-wavelength radiation most. As a result, the scattered radiation coming to us from all parts of the atmosphere gives the clear sky its characteristic blue colour when seen from low altitudes. Droplets of water also scatter radiation strongly, and where they occur densely, as in a thick cloud, multiple scattering may cause as much as 80% of the incident photons to be turned back out into space. Since the earth's average cloud cover is about 50%, this is a powerful mechanism for loss of solar energy.

True absorption by molecules of water vapour, carbon dioxide and other constituents of the atmosphere is another important barrier to the incoming rays. It is wavelength-dependent, with a number of strong absorption bands, occurring mainly within the infra-red region. All electromagnetic radiation travels with the characteristic velocity, c, about 300 000 km (186 000 miles) per second. The frequency of waves of wavelength λ is thus c/λ, and for the infra-red component of the sun's radiation, the frequency has values corresponding to the frequency of bond vibrations in some of the molecules constituting the atmosphere. Energy can be taken up from the waves in these vibrations and is distributed to other molecules through the frequent occurrence of collisions in our relatively dense atmosphere. For our present purposes we might consider absorbed radiation to have been lost, though we shall see later that the re-emission of the energy has important effects.

These interactions with the atmosphere reduce the intensity of solar radiation at the surface to little more than half the value at the top of the atmosphere. They also bring about changes in the spectral distribution of energy, by absorption, and in the direction from which energy arrives at the surface, by scattering. These effects depend upon the local constitution of the atmosphere and vary appreciably from place to place. Pollution near centres of population, high water vapour content near coasts, and changing patterns of cloud cover, all serve to make prediction of energy intensities difficult. Perhaps the most important factor is the length of the path of the sun's radiation through the atmosphere in reaching a particular location. This determines the losses through scattering and absorption, in particular. It varies in a rather complicated way with time of day, season of year and position on the earth's surface, as we shall now see.

2.4 *The apparent position of the sun*

The most important features of the very complicated motion of the earth are its passage round the sun once a year and its rotation once a day. Although the earth's orbit is strictly an ellipse, the eccentricity is so small that it is barely distinguishable from a circle. The mean radius is about 150 million km (92·9 million miles), defining the astronomical unit of distance; the extreme variations from this during the year amount to some $1\frac{1}{2}\%$ only. This variation is not noticeable and plays no significant part in the earth's changing seasons. These are caused by the inclination of the axis of rotation, which is not perpendicular to the plane of the orbit. The inclination relative to that plane is about $66\frac{1}{2}°$, and, since there are no forces acting to move it, the axis points in a direction which is essentially fixed in space.*

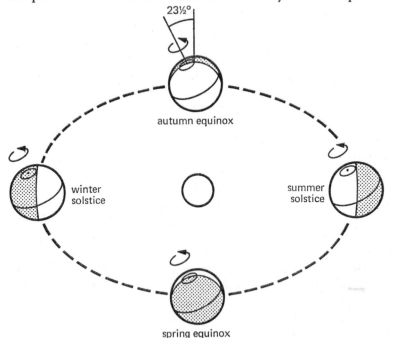

Fig.2.3 Motion of earth relative to sun

*This direction is, in fact, slowly changing, due to a free precession with a period of about 26000 years.

As may be seen from Figure 2.3, this causes substantial variations in the duration of daylight during the year, as the earth moves round its orbit. A point in the northern hemisphere will have its longest period of daylight at the time of the SUMMER SOLSTICE, at present June 22nd, when the northern end of the axis points towards the sun, and its longest period of darkness at the WINTER SOLSTICE, on December 22nd, when the axis points away. The two neutral points in between, when the earth's axis is perpendicular to the line joining the earth and the sun, are the EQUINOXES, coming round March 21st and September 23rd. On these dates, day and night are of equal length.

The apparent motion of the sun, as seen from a point at latitude L in the northern hemisphere, is as shown in Figure 2.4. We define the sun's apparent position from any point on earth by two angles, the ALTITUDE A relative to the horizontal and the AZIMUTH Z relative to due south. When the sun is due south, the azimuth is zero and the altitude a maximum. At this

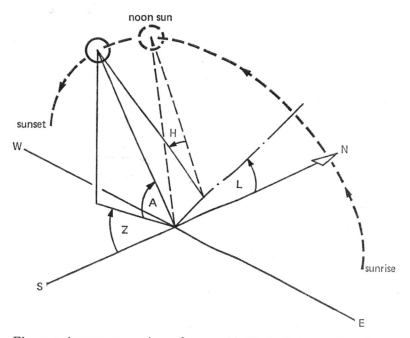

Fig. 2.4 Apparent motion of sun at latitude L in northern hemisphere

instant it is said to be SOLAR NOON, and it is a convenient origin from which to measure time of day.

Because of the earth's rotation, the sun appears to us to revolve, once in 24 hours, round the earth's axis. The earth being small in relation to the distance of the sun, we shall not introduce much error if we translate this axis so that it passes through our location at latitude L, as shown. The axis is evidently inclined at an angle L to the N–S horizontal at that point. If we measure the sun's hourly motion relative to solar noon, it will have progressed round the axis by an angle H, known as the HOUR ANGLE, at a time, t hours after noon, given by

$$H = \frac{t}{24} \cdot 360° = 15t \text{ degrees,}$$

or $$H = \frac{t}{24} \cdot 2\pi = 0.252\ t \text{ radians,} \qquad 2.1$$

since 24 hours are required to encompass 360° or 2π radians. Seasonal variation in the sun's apparent position is represented by the angle of DECLINATION, D. This varies between $+ 23\frac{1}{2}°$ at the summer solstice and $- 23\frac{1}{2}°$ at the winter solstice, as can readily be deduced by consideration of the motion shown in Figure 2.3. The motion depicted in Figure 2.4 is evidently taking place in the summer, for there D is positive, with the sun taking a path which intersects the horizon north of the E–W line at sunrise and sunset. At the equinoxes, D is zero, so that the sun rises and sets exactly on the E–W line. The solar declination at other times of the year is best obtained from tables, though for most purposes a rough calculation is permissible. If we measure the time of year in days d from the spring equinox, the declination is given approximately by

$$D = 23\frac{1}{2} \sin \left(\frac{2\pi d}{365}\right) \text{ degrees,} \qquad 2.2$$

since the sine function varies between $+ 1$ and $- 1$ in the way required.

We are now in a position to calculate the apparent position of the sun at a point with latitude L at a time of day represented by the hour angle H and season of year represented by the declination D. At solar noon, the altitude evidently has its maximum value of $90° - L + D$, but at other times the determination of the position is a little more difficult. The reader who enjoys a tussle with trigonometry will find this an interesting exercise, though charts and tables are available. We shall merely state the results here, the simplest forms being that

$$\sin A = \cos D \cos H \cos L + \sin D \sin L, \qquad 2.3$$

and $$\sin Z = \frac{\cos D \sin H}{\cos A}. \qquad 2.4$$

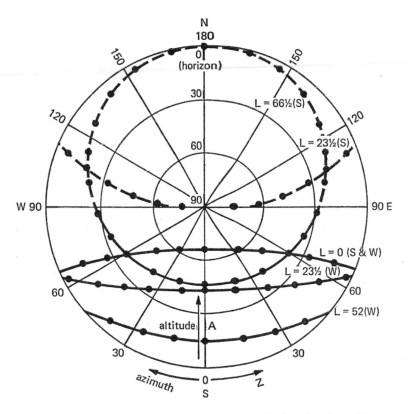

dots indicate hourly positions
relative to solar noon

Fig. 2.5 Apparent path of sun at various latitudes and seasons (northern hemisphere)

There is no completely satisfactory way of representing the three-dimensional apparent motion of the sun in a diagram, but the method adopted in Figure 2.5. brings out many of the essential features. That diagram shows the angular co-ordinates of the sun's apparent position, A and Z, as functions of time relative to local noon. Paths are given for representative positions on the earth at various times of the year; the equator (lat 0°), tropics (lat $23\frac{1}{2}$°), northern Europe (lat 52°) and arctic circle (lat $66\frac{1}{2}$°). Those parts of the paths for which the azimuth exceeds 90°— that is, when the sun is on the north side of the E–W plane— are shown dotted. We may note here the phenomenon of the 'midnight sun' at and above the arctic circle, when the sun is visible all the time in summer. British readers may note the very low sun path in Britain at mid-winter.

Sunrise and sunset occur when $A = 0$. From equation 2.3, this will occur at hour angles H_1 either side of noon given by

$$\cos D \cos H_1 \cos L \ = \ - \sin D \sin L,$$

that is, by $$\cos H_1 = - \tan D \tan L. \qquad 2.5$$

For example, in the United Kingdom ($L = 52$°) we have $H_1 = 124$° (8·3 hours at 15° per hour) before and after noon in midsummer ($D = 23\frac{1}{2}$°) and $H_1 = 56$° (3·7 hours) in mid-winter ($D = -23\frac{1}{2}$°). The increase in hours of daylight in summer is already noticeable at the tropics ($L = 23\frac{1}{2}$°) where, at the midsummer solstice, the sun is above the horizon for about $13\frac{1}{2}$ hours. We may see from Figure 2.5 that at this time of year the sun is directly overhead at noon and has an azimuth close to 90° all day. These characteristics have an important effect on the tropical climate, as we shall see later.

2.5 Intensity on a surface
Two factors strongly influencing the intensity of solar radiation falling directly on a surface are the obliquity of the rays relative to the plane of the surface and the length of the path of the rays through the atmosphere. Both these factors depend upon the apparent solar altitude A.

If, as shown in Figure 2.6, rays of intensity I per unit area fall upon a plane surface in a direction inclined at an angle θ to the NORMAL, or perpendicular, through the surface, the

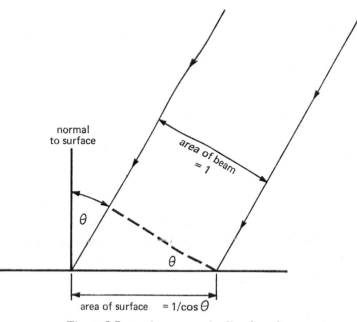

Fig. 2.6 Intensity on an inclined surface

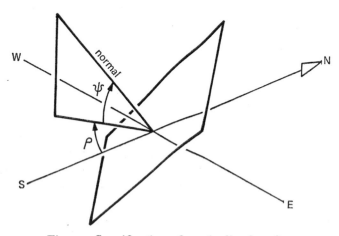

Fig. 2.7 Specification of an inclined surface

area covered by unit area of beam is $1/\cos\theta$. Then the intensity on that surface is $I\cos\theta$. We can represent the inclination of any surface, such as a hillside slope or the roof of a house, by specifying the direction of its normal in the same way as we specified the apparent position of the sun: by assigning an altitude ψ and aximuth φ. These angles are shown in Figure 2.7. Though we need not go deeply into the determination of intensities for such inclined surfaces, we should have an outline of the method for completeness. Some further manipulation of the trigonometry shows that the angle θ between the sun's direct rays and the normal of such a plane is given by

$$\cos\theta = \cos(A-\psi) + \cos A \cos\psi [\cos(Z-\varphi) - 1]. \quad 2.6$$

We can see here that for surfaces with $\varphi = Z$, whose normals lie in the same vertical plane as the sun, the angle θ is equal to $(A - 90°)$. Then $\cos\theta$ is equal to $\sin A$ in that important case.

The length of the direct ray's path through the atmosphere is almost inversely proportional to $\sin A$, as may be seen from Figure 2.8, because the earth's atmosphere is so thin compared with its radius. Only for very low solar altitudes does the curvature of the earth become important in this. This is a very useful simplification. Meteorologists relate the length of the path to the AIR MASS such that a direct radial path corresponds to air mass $= 1$. Then for any altitude A, the air mass is equal to $1/\sin A$. The idea here is that the scattering and absorption experienced by a solar ray is proportional to the mass of atmospheric constituents encountered. For example, when A has the values 42°, 30° and 20°, the air mass is $1\frac{1}{2}$, 2 and 3 respectively. Thus, in northern Europe or central Canada in midwinter, the maximum solar altitude is less than 15°, corresponding to an air mass of 4, so that even at noon the sun's rays have to traverse a path equivalent to four vertical journeys through the atmosphere.

The reduction of the intensity and modification of the spectral distribution of solar energy by absorption and scattering depend in a fairly complicated way upon the length of the air path, or the air mass. The effects are shown approximately for cloudless atmospheres in Figure 2.9. The general reduction is largely due to scattering and the deep fissures to absorption by water vapour and carbon dioxide. The areas under these curves represent the intensity I, the total energy per unit time per unit

36

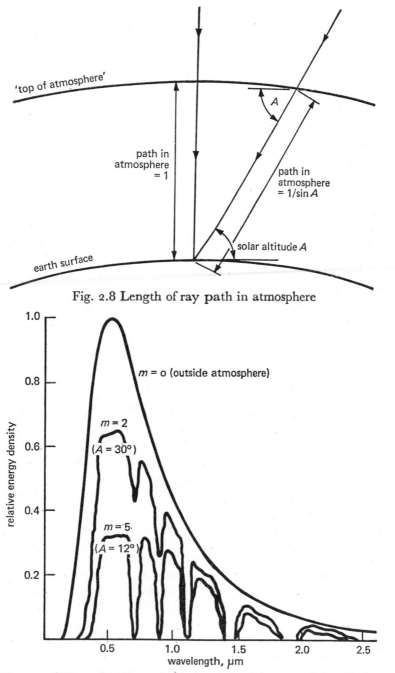

Fig. 2.8 Length of ray path in atmosphere

Fig. 2.9 Effect of passage through atmosphere on solar radiation

area on a surface directly facing the sun. For solar altitudes of 90°, 30°, 20° and 12° respectively, (corresponding to air masses of 1, 2, 3, and 5) the intensity I is about 900, 750, 600 and 400 watts per square metre in a clear atmosphere. Approximate values for intermediate altitudes may be obtained by inter-polation, between these values, according to the appropriate air mass value. (We note here that since the intensity is the *rate* of passage of energy per unit area, it represents *power* and is given in an appropriate power unit such as the watt per m^2.)

There are minor changes also in the distribution of the energy with wavelength at different locations, times and seasons. On the average, we find only a trifle, say 1%, with a wavelength less than $0.35\mu m$, about 50% in the visible band ($0.35-0.75\mu m$) and about 50% in the infra-red range.

The total energy falling upon a surface is greater than the values given above. It consists not only of the DIRECT component just considered, but also of a DIFFUSE component, comprising radiation which has been scattered by constituents of the atmosphere and re-directed onto the surface, as seen in Figure 2.2. The diffuse radiation, sometimes called the SKY-LIGHT, represents a significant fraction of the total, particularly at very low solar altitudes, where it can be as much as a half. It is difficult to estimate with much precision, and numerous different values have been given by various authorities. It is known, however, to depend primarily on the prevailing alti-tude of the sun, and for our rough calculations here, we shall take some approximate values for different solar altitudes corresponding, as for the direct component, to clear conditions with air masses of 1, 2, 3 and 5. For these conditions, the diffuse radiation intensity I_d on a *horizontal* surface is about 110, 90, 70 and 50 watts per square metre respectively. We shall interpolate between these values as required. When dealing with sloping surfaces, we shall not be justified in attempting to allow pre-cisely for the variation of the diffuse radiation with direction. Here we shall simply reduce the values for horizontal surfaces by a factor which decreases linearly with slope from 1·0 for horizontal surfaces to 0·5 for vertical surfaces (which can only 'see' half the sky). The energy distribution of the diffuse radia-tion is shifted slightly towards the shorter wavelengths com-pared with that of the direct radiation, largely because of the stronger scattering by atmospheric constituents at this end of the spectrum.

2.6 *Daily and yearly radiation*

To make a first assessment of the economic value of solar radiation, we shall need to estimate the total energy received on a surface in a given time, say a day or a year. To do so, we must add up the energy received directly along a path whose direction and associated air mass change with time, together with the diffuse contribution. The accuracy with which these calculations can be made is greatly affected by variations in the atmospheric conditions, through pollution, fog, ground haze, cloud cover and so on. These phenomena serve, in general, to reduce the direct component by further absorption and scattering, but the scattering may enhance the diffuse component considerably. With complete overcast cloud cover, for instance, *all* the radiation reaching the ground is diffuse.

The best way of obtaining the values we require would be actually to measure the intensity over long periods of time at many places. Already, a number of meteorological stations throughout the world keep a continuous record of the intensity on a horizontal surface, a quantity known, somewhat infelicitously, as the INSOLATION. The number of centres for which data are available is likely to grow as awareness of their importance increases, though at the moment, they are too few to give a reliable picture for the world as a whole.

We can, however, make some rough calculations for present purposes. Since any exploitation of solar energy on a large scale will occur first where the intensity is greatest, we will consider only locations which enjoy a clear atmosphere. We have already, in previous sections, the material with which to estimate the insolation for a horizontal surface in any part of the world. At any time represented by the hour angle H, the solar altitude A is given by equation 2.3. The intensity I of the direct radiation can then be estimated, from its dependence on the air mass, equivalent to $1/\sin A$, as given in the preceding section. For a given surface, the intensity is then $I \cos \theta$, which is simply $I \sin A$ when that surface is horizontal. The additional contribution for the diffuse radiation is merely the intensity I_d, which can also be estimated once the solar altitude A is known.

The variation of the insolation with time, relative to solar noon, for a latitude corresponding to central Britain $(L = 52°)$ is shown as an example in Figure 2.10. The two curves given are for the summer and winter solstices. The area underneath the curve represents the total energy received throughout one day and the two values given represent the maximum and minimum for the year; as can be seen, the variation at this latitude is considerable.

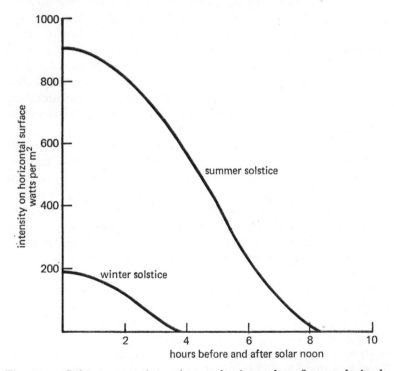

Fig. 2.10 Solar energy intensity on horizontal surface at latitude 52°N (clear atmosphere)

Similar calculations for other representative latitudes lead to the values given in Table 2.1. Two values are given for each; the second (in parentheses) including the diffuse contribution, the first being the direct component only. The unit of energy employed in this table is the kilowatt-hour, the familiar 'unit' of our domestic electricity supply.

TABLE 2.1
Insolation at various latitudes for clear atmospheres

Location	Latitude	Insolation kWh/m² Maximum	Minimum	Yearly Total
Equator	0°	6·5 (7·5)	5·8 (6·8)	2200 (2300)
Tropics	23½°	7·1 (8·3)	3·4 (4·2)	1900 (2300)
Mid-earth	45°	7·2 (8·5)	1·2 (1·7)	1500 (1900)
Central U.K.	52°	7·0 (8·4)	0·5 (0·8)	1400 (1700)
Polar Circle	66½°	6·5 (7·9)	0 (0)	1200 (1400)

We might note, in passing, that in the case of the equator, the maximum insolation occurs at the equinoxes, when the sun's azimuth is 90° all day, and its path passes overhead. At that latitude, the *minimum* insolation occurs at mid-summer and mid-winter; it is only beyond the tropics that the more familiar pattern occurs.

To obtain the yearly totals, we have to integrate again, adding the contributions from each day throughout the year. The variation of insolation during this time is roughly similar to a sine wave between the maximum and minimum values, with a symmetrical distribution in the two halves of the year. Approximate values of the yearly totals of energy received by a horizontal surface in clear conditions are given in the final column of Table 2.1.

These results bring out the little-known fact that the highest daily insolation values do not occur at the equator, but around latitude 40°. This is largely a consequence of the sun's maximum declination, due to the inclination of the earth's axis. At mid-summer in the tropics, the sun's path is almost directly overhead, as seen in Figure 2.5, and the hours of daylight are greater than at the equator (about 13½ hours). The lengthening day as latitude increases largely compensates for the diminishing solar intensity, so that the midsummer values have a maximum around 40°, but are almost constant, for clear conditions, up to the polar circles. On the other hand, the midwinter values fall sharply with increasing latitude, and the yearly total at the polar circles is only half that at the equator.

It must be emphasised that the values given in Table 2.1 are for clear conditions only. Overcast cloud, or a combination of broken cloud cover and industrial pollution, common in many countries, can bring about reductions of the order of half or more of the values given. In Britain, for instance, the actual annual insolation is only some 900 kWh/m², rather than the clear-air value of 1700 kWh/m² for that latitude.

2.7 The promise

The solar energy falling on the roof of a dwelling in a year far exceeds the energy used in the most highly-developed countries for heating or cooling such a dwelling. It is natural to think first of using solar energy for purposes such as this, and in Chapter 5 we shall be looking at practical methods of doing so. These, and other straightforward uses of the sun's energy, such as the drying of foodstuffs and the distillation of water, are already well advanced in several countries. We have mentioned in Chapter 1 a number of requirements, which could play a major part in the re-vitalisation of the rural economy in under-developed countries. For many of these, the sun's energy would have to be converted into electricity or mechanical work. Chapters 6 to 9 will deal with possible energy-converting devices for these purposes. We shall find that most of these devices are very inefficient, so that only a fraction, perhaps a few per cent, of the energy collected can actually be furnished for a human purpose. Nevertheless, the possibilities seem exciting. With a recovery rate of 5%, the present world-average energy demand per person could be met by the solar energy falling on a horizontal surface a mere six metres square in the tropics. Even the probable long-term future demand would represent only 14 metres square per person. For many years, prophets have been pointing out that the earth has ample wasted space which could be utilised in this way, without conflicting with other needs, such as the production of food.

The favourite sites proposed for the power stations of the solar millenium are the great desert belts which almost ring the earth at the tropics. These vast parched areas, shown in Figure 2.11, occur deep within the large land masses, particularly where the prevailing winds carry little moisture. This happens when the air has been cooled by passing over a cold ocean or by having risen over a mountain range. In such regions, clouds are almost non-existent and rain comes only at irregular intervals as a result of random changes in the wind structure. The driest areas, with less than 25 cm of rainfall a year, occupy nearly an eighth of the world's land surface. These are the true deserts, lying along the tropics on either side of the equator. Round the Tropic of Cancer, north of the equator, are the Great Western Desert of North America, the Thar Desert of India, the Arabian Desert and the Sahara; south of the equator, along the Tropic of Capricorn, lie the Atacama Desert of South America, the Kalahari Desert of South Africa and the Great Victoria Desert and others, that occupy nearly half of Australia. These arid wastes produce nothing and nourish no one. Their total area is some 20 million square kilometres (7m square miles). On them falls about 50 000 million million kWh of energy a year. At 5% recovery, this would meet a demand nearly two hundred times as great as our present needs. (It is salutary to note, however, that it would meet only twelve times our probable future needs, as estimated in Chapter 1).

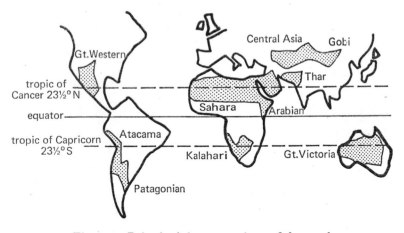

Fig. 2.11 Principal desert regions of the earth

Intriguing as these figures may be, the prospect of recovering solar energy on this scale can be entirely discounted. The low intensity of solar radiation makes the provision of suitable collectors quite impracticable. An area 3 km square would be required to produce power at the rate of a single modern power station. To cover a sizeable part of the world's deserts, with the simplest imaginable collecting structure, would require more than our known reserves of metals.

We would be wiser to have more moderate aims. Solar energy falls in other places, beside the deserts. In fact, it falls in useful amounts where it is needed most. About 80% of the world's population lives between the 40° parallels. Most of them can count on an insolation of at least 1500 kWh/m² per year. It would be more useful to compare this, not with the 3000 million kWh per year produced by a European power station, but with the 1000 kWh per year of work performed by a pair of bullocks, and the cost of recovering solar energy with the cost of the 25 000 kg of fodder per year that they consume.

We shall not be able to take these economic comparisons far in the present book. The main purpose of the ensuing chapters is to review the scientific basis of a number of methods of recovering solar energy for human purposes. This will provide a broad base upon which such comparisons can be founded. Before beginning this, there follows an interlude in which we examine some of the fundamental concepts from which the scientific development springs.

3

A REVIEW OF FIRST PRINCIPLES

Sweet is by convention and bitter by convention, heat by convention, cold by convention, colour by convention; in reality there are only atoms and space.

DEMOCRITUS OF ABDERA (*c*. 460–*c*. 370 B.C.)

THROUGHOUT this book we shall need to use certain concepts repeatedly and it is important to begin by having a clear understanding of what they mean. Already we have used terms like energy, heat, temperature and work, which are part of our everyday vocabulary. But from a scientific point of view, they have much more restricted meanings than those we normally attach to them. Moreover, in our common usage, we are generally unaware that concepts such as these are often difficult to define because they are inter-related, so that it is instructive to examine the foundations from which they are derived.

Terms such as these are part of the fundamental language of THERMODYNAMICS. This is not intended to be a textbook on the subject or even an introduction to it, so that we will not need to consider all of even its primary terms. However, in this chapter we shall take an initial look at the basis of some of those thermodynamic concepts which will be used in later chapters. This will be a digression from the main theme of the book and will be kept fairly short so that the development will not be too much delayed.

Further concepts will be examined later in the book when the need to introduce them arises. If we are to deal with the many possible ways in which solar energy could be used to human advantage, our study will lead us across many of the traditional boundaries between the scientific disciplines. We shall have to tread in what may be for many the unfamiliar territory of quantum mechanics and radiation chemistry. This is nothing new to the engineer, who spends his life ignoring boundaries. He finds that he must often gain a working understanding of

45

fields which do not usually figure in his curriculum. He is helped to do this by an attitude of mind which relates everything to what are called FIRST PRINCIPLES. These are the common framework of ideas which form the skeleton of all the sciences. We shall only discuss here the parts of the framework which we need. Sometimes, we shall have to take things for granted, and we can only do this justly if we can be confident that they do not violate the first principles. To do that, we shall have to be sure what these principles mean. For the most part, we shall find that what we need here can be related to the DYNAMICS OF PARTICLES, which forms part of the physics encountered in an ordinary secondary education. This, which is the foundation for so much else, is part of the immeasurable legacy of ISAAC NEWTON (1642–1727).

3.1 *Reality and the conceptual model*
We shall have to deal here, as in much of science, with things of which we cannot have direct experience. We commonly speak of atoms and molecules, electrons and waves, and so on, which are not accessible to our unaided senses, nor can they be made discernible to us by the most elaborate and sensitive apparatus. The existence of something which we call an electron, for example, has been inferred because a multitude of reproducible phenomena can only be explained by supposing such an entity, having certain properties, to be present. But we cannot see an electron and it is natural, and sometimes helpful, to try to visualise it. We try to imagine something, similar to things within our experience, which could have properties which the electron is known to exhibit. This is a CONCEPTUAL MODEL, in terms of which we can discuss the probable behaviour of electrons in various circumstances. Most scientific analysis takes place in terms of models of this kind.

For instance, the behaviour of electrons in many experiments shows that each has a certain mass, a fairly well-defined size and other properties which we associate with material bodies. It is natural then to think of it as a classical particle; this is our first model of the electron. The motion of particles which are big enough for us to see is described quite precisely by the dynamical principles known as Newton's laws of motion (and, in certain circumstances, the refinements of these due to Einstein). These laws are part of our first principles. In several centuries of experience with dynamics we have found no exceptions to them. If we apply these laws to predict how an electron would behave, we find that the predictions are, in general, correct. Then our particle model of the electron is a useful one.

But this does not mean that an electron *is* a particle. We find that in some other circumstances, it exhibits properties that a particle could not have. Certain diffraction experiments, for example, show that an electron must have passed simultaneously through two adjacent apertures (actually the spaces between atoms in a crystal) further apart than the size which would be attributed to it through its behaviour elsewhere. In this, and in its behaviour after passing through the apertures, the electron has properties which are characteristics of waves, such as we find with light. It is now known that all small things show this DUALITY of behaviour, and we just have to get accustomed to it. Thus, an electron sometimes shows the properties of a particle, sometimes those of a wave. We cannot imagine any single thing with characteristics like these, so we cannot use a single model to describe the electron in every circumstance.

It was said above that many of the activities of the electron are described perfectly adequately by supposing it to be a particle. If we did not need to discuss circumstances in which the particle model is inadequate, we would come to no harm by using this model always. In fact, there is much to be said for using the simplest model whose characteristics are known to represent the reality as nearly as we need in any given situation. That is what we shall do here. We need not be concerned that some of our models would not be suitable for more refined analysis. But we should beware that usage and familiarity do not lead us to think that our model is in any sense real. It is meaningless to ask, for example, what an electron *is*. In using a particle model to represent it, we are not saying that it *is* a particle, but that for our present purposes, we can adequately describe its properties by supposing that it *behaves* like a particle. There is another feature of conceptual models which is also of interest from a philosophical viewpoint. Sometimes it is possible to discard an inadequate model, once its inadequacy has been recognised, but to retain much of the advance in understanding which could only have been obtained by using this model. An outstanding example of this is the development of the notion of ENERGY LEVELS through the use of models which have long since been discarded. This process will be illustrated as our study develops, for the concept of energy levels is central to much of the later work of the book.

In the following sections, brief accounts are given of those models upon which much of the reasoning elsewhere in the book is based. These models are as simple as possible, with only enough complexity to serve our immediate needs, though they are also adequate for many other purposes.

3.2 *Structure of matter*

All matter is considered to be made up of elemental ATOMS. Each atom has a NUCLEUS, consisting of a tightly-knit assembly of relatively massive particles, NEUTRONS and PROTONS, giving it a positive ELECTRIC CHARGE. The atom as a whole has no nett charge, because of the ELECTRONS, each of which has a relatively small mass but negative charge, orbiting the nucleus. It will be sufficient for our purposes to use the conceptual model of the atom suggested by BOHR (1913). It is found that only certain orbits are permitted to the electrons and these orbits can be considered to be arranged in a regular sequence, forming SHELLS surrounding the nucleus. There is a maximum number of electrons which a given shell can contain, though in many atoms it is found that the shells do not contain their full complement. Chemical combination of atoms into MOLECULES takes place through the inter-linking of the outer shells and the number of electrons in these shells determines which elements can combine together and the proportions in which they do so. The measure assigned to the propensity to combine is called the VALENCY and the outer electrons are accordingly termed VALENCE ELECTRONS. In the case of electrical conductors, particularly the metals, some of the electrons are so far from the nucleus and so weakly held that an individual electron cannot be said to 'belong' to any particular atom or molecule. At any instant there are always enough electrons present in a body to satisfy the condition of zero nett charge for the whole body, but the electrons are fairly free to move about within it.

When a body of atoms or molecules come together in a group, they are affected by forces, of electrostatic origin, acting between

48

each pair. It is conjectured that if two atoms were at rest, the nett force exerted by each on the other would vary with the distance between their centres in the manner shown in Figure 3.1. This nett force appears to be a combination of two actions, one a strong repulsion which falls off very rapidly with distance and the other an attraction, also falling with distance, but which is effective over longer distances. As two atoms are brought closer together, the nett force is at first attractive, increasing as the distance decreases until the point A is reached, at which the repulsion begins to dominate. Thereafter, the nett force changes rapidly to a repulsion, passing through a point B at which it is zero. Two atoms at rest would remain at this distance apart, since an external force of one sort or another would be necessary to push them closer or pull them further apart.

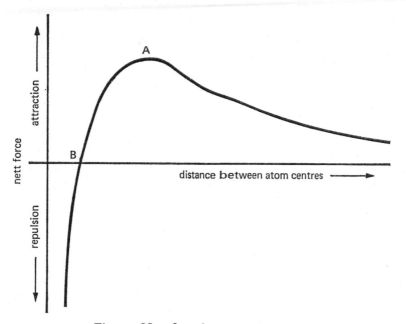

Fig. 3.1 Nett force between two atoms

In reality, atoms and molecules are never at rest. In a solid, where they are mostly arranged in a regular pattern called a LATTICE, they remain close to an equilibrium point such as B in relation to each other (though this is affected by the other near neighbours) but they vibrate relative to it. As can be seen from Figure 3.1, the nett repulsion to one side of B increases much more rapidly with distance than does the nett attraction to the other side. Thus, when the atoms are vibrating, more of the motion occurs on the 'weaker' side and the centre of the oscillation is further out than B. As we shall see later, increases in the range, or AMPLITUDE, of these oscillations is associated with an increase in temperature. Thus, this increase in average separation between the atoms is manifested in the phenomenon of expansion on heating.

Further increases in amplitude of oscillation result in a breakdown in the regular arrangement of relative positions, typical of the SOLID STATE, and a looser arrangement, which we call the LIQUID STATE, results. Here, the average distance between atoms or molecules is greater than in the solid state, but they are still strongly influenced by the forces that act between them. In the GASEOUS STATE, each molecule has become sufficiently energetic to be able to move about at random, without being seriously influenced by the others. They are able to approach and recede from each other without entering into permanent associations, and can spread out to fill any container into which the assembly is placed. Nevertheless, their frequent approaches, amounting almost to collisions, ensure that at any instant their velocities are not very different from each other, although some will always be moving momentarily more slowly or more quickly than the average.

3.3 Energy

Energy is one of the key concepts of science, but it has been acquired gradually and the range of different situations in which it is now found to be useful precludes a brief definition.

We can begin by considering WORK. This is a quantity of which we have direct experience through physical effort. We define it formally in terms of the movement of something under the action of a FORCE. If a force P is applied to a body, which moves a distance x in the direction of action of P, we define the work done on the body as the product Px. Now this work can affect the body in several different ways. If P is applied to an unrestrained body, for example, the velocity v of the body will increase; if forces are applied to a deformable body such as a spring, the shape of the body will change. In simple cases like these we say that the ENERGY of the body has increased by an amount equal to the work done on it. For the moment, if we neglect more complex systems involving chemical reactions and the like, we can say that, in general, any increase in energy of a system undergoing a change in its state is equal to the work that would have to be done on it to bring about such a change.

The several ways in which a body can store energy are distinguished by different names, though we need concern ourselves here only with the two most common: KINETIC and POTENTIAL energy. If a particle of mass m is to be given a velocity v, by a gradual acceleration from rest, it is easy to show that the work that must be done on it is equal to $\frac{1}{2}mv^2$; this is then its increase in kinetic energy or energy of *motion*. When a spring is slowly compressed, the work is said to have produced an increase in its potential energy or energy of *position* which can be expressed in terms of its change in shape.

We can identify these energy terms in the behaviour of atoms and molecules. In a gas, the molecules are moving about, so they have kinetic energy by virtue of this TRANSLATIONAL motion. Work also has to be done to increase the spin of rotating things, so that molecules which are spinning also possess kinetic energy by virtue of their ROTATIONAL motion. In the case of the single atoms of a monatomic gas, the rotational energy is negligible but when they are grouped into molecules of considerably greater size than a single atom, the rotational part of the kinetic energy can be important. In a molecule, too, the interatomic forces act somewhat like springs between the atoms, so the system has potential energy as well. When the atoms oscillate relative to each other, there is an interchange of kinetic and potential energy. The motion of the molecules of a solid can be visualised in the same way, though here the behaviour is much more complex.

Within the atom, the electrons can be said to possess potential and kinetic energy because of the attractive forces between them and the positively-charged nucleus. The electrons can be imagined to revolve about the nucleus along a path in which this attractive force provides the necessary centripetal acceleration. Work would have to be done to move an electron out into a more distant orbit, so this will involve an increase in its energy. We shall return to this situation again in Section 4.1.

Our examples have involved energy associated with purely *local* movements of molecules and their constituent particles. This type of energy will be called INTERNAL ENERGY. This is to distinguish it from the energy associated with the *gross* movements of the bodies to which the molecules belong—when a solid body is in motion or a liquid flowing in a pipe, for example. In these cases, the motion is a combination of a large-scale movement superimposed upon the local ones but it is convenient to think of these components separately.

3.4 *Heat and Work*

As we have seen, the internal energy of a given body can be increased by doing work on it. In the case of a body of gas in a vessel, we could do this by stirring it vigorously or by compressing it. Alternatively, we could increase the energy by *heating* the gas. We know that the water in a kettle can be turned into steam by heating. The change from the liquid to the gaseous state involves an increase in the average distance between the water molecules and their motion must be more energetic to achieve this. In all the instances given, what has taken place is a transfer of energy to the system from elsewhere. However, there are reasons, which will appear later, for continuing to identify the agency by which this transfer is effected either as HEAT or as WORK. Although we have spoken so far of increasing the internal energy by these agencies, we should note that the internal energy can be *decreased* also by the transfer of energy *from* the system to the surroundings. When this happens, the transferred energy can again be identified as heat or as work.

We distinguish between these agencies by the manner in which the energy transfer takes place. If it is through a change in position of one of the boundaries of the system, as when a gas is compressed in a cylinder by the movement of a piston, the energy is said to be transferred as work. Heat, however, is energy being transferred as a result of a difference in TEMPERATURE between the substance and its surroundings. This is a new concept and we must next examine it in some detail, before returning later to a further consideration of energy transfer.

3.5 *Temperature and the Zeroth Law*

We all have a primitive concept of hotness and coldness through our sense of touch. Though this sense is rather imprecise, several observers would by this means assign the same place to each of a series of similar bodies if asked to put them in order of increasing hotness. Here they are having to decide the *relative* hotness of bodies. But they would not agree with each other about whether a given body can be described as 'hot', 'warm', 'cool' or 'cold', because this would require an *absolute* judgement to be made. The body would then have been placed somewhere on a crude SCALE of hotness; this would only be possible if there had been some standard against which hotness could be judged. When such a scale has been established, we speak of the TEMPERATURE as an agreed measure of hotness, using the word mainly to mean an absolute numerical measure of this property. We shall return shortly to the question of establishing a scale of temperature, but for the moment we can continue to think of it only in the sense in which we say that the hotter body has a higher temperature than the colder one.

Now we know that in general an increase in temperature of an isolated body is associated with an increase in its internal energy. The atoms and molecules of which it is composed move, spin and oscillate more vigorously as the temperature rises. We cannot see these motions directly, but we are satisfied that they occur because we can observe a multitude of otherwise unconnected phenomena that would logically be expected to follow if they did occur.

When two solid bodies at different temperatures are brought together, interactions occur between the atoms and molecules of the contacting surfaces. As a result of these interactions, the vibrations of the less vigorous ones of the cooler body are increased and the vibrations of the more vigorous ones of the hotter body are diminished. In fact, at this local level, work is done through the action of the intermolecular forces. There has thus been a transfer of energy from the hotter to the cooler body. Such a transfer, *brought about by a difference in temperature*, is called HEAT TRANSFER, though only as the energy is in transit do we identify it as heat.

The effect of this heat transfer on bodies of various kinds will depend upon their constitutions. If the system is a body of gas, it merely increases the average kinetic energy of the molecules. The interaction of these with the molecules of the walls of the containing vessel, which are in effect fixed in position, is rather like the rebound of particles from a fixed surface on a bigger scale. With an increase in energy, this rebound is more violent, and we shall say that the pressure has increased. In a solid, the atoms will vibrate through greater amplitudes, and their mean positions will move further apart. There are so many atoms in a solid that the overall change in dimensions will be easily measurable, and we shall say that the body has expanded. More complicated behaviour, such as melting or evaporation, may ensue, though this is also mainly explicable in terms of straightforward mechanics. All that we need note for the moment is that the temperature of different bodies cannot be expected to change by the same amount when a given amount of energy is introduced. The change will depend on how many particles have to be affected, how massive they are, how many ways they can take up energy, and so on.

When two bodies at different temperatures are placed together, heat transfer occurs until a state of EQUILIBRIUM is reached. In this state, interactions between the atoms bring about no further changes in any measurable property. We have seen that there can be no particular relation between the internal energies of the two bodies in this state. But we shall expect from our common experience that they have the same temperature, taken to be a measure of the property of hotness. This expectation is based on observation, and is not dependent on how this property is to be measured. Such observations show that any third body, brought into equilibrium with one of the original two, will also be found to be in equilibrium with the other, if they are subsequently brought into contact. This is a statement about the way things happen in our particular universe, and is called the ZEROTH LAW OF THERMODYNAMICS. It has this unusual title because it was not recognised until after the establishment of the first and second laws (which we shall consider later) though it is now seen that it must logically precede these.

If, then, we can have any number of bodies, of any conceivable kind, different in shape, size and constitution, simultaneously in thermal equilibrium, what do they have in common?

One of these bodies could have been a device for giving an agreed indication of temperature. Evidently, it will give the same indication whichever of these bodies it is in contact with, since we have defined equilibrium as a state in which no measurable changes occur. Thus, they are all at the same temperature. This corollary of the zeroth law was first given by JAMES CLERK MAXWELL in 1868.

3.6 Scales of temperature and the ideal gas

We have come now to a first definition of temperature, adequate for our present purposes. It is a property which has the same value for all bodies which are in thermal equilibrium with each other. But we need a standard way of assigning a value to the temperature if we are to make calculations about the thermal behaviour of bodies. For this, we must have a device which can be reproduced anywhere, and which has some easily measurable property which varies continuously with temperature in a regular way. Such a device is called a THERMOMETER. Once brought into contact with the body whose temperature is required, it will come into equilibrium with it and we can give a value to the common temperature in terms of the measurable quantity—say the length of a mercury column or the electrical resistance of a wire. The difficulty is that all measurable quantities will not depend on temperature in the same way, so we must agree on a standard scale and relate the others to that.

One such standard, which has been widely used, is the IDEAL-GAS SCALE OF TEMPERATURE. This is based on the notion of an ideal gas, whose only capacity for storing internal energy is as translational kinetic energy of its molecules. Then the temperature of such a gas is a simple function of its internal energy. We can obtain the characteristics of the ideal gas by employing elementary mechanics, that is, we use what is called the KINETIC THEORY OF GASES. The systematic formulation of this theory in recent times is generally attributed to the German physicist RUDOLF CLAUSIUS (1857), though the ideas involved had been current for at least a century previously. The essential features of the model on which the theory is based are that the molecules of a gas can be assumed to behave like classical particles, to move independently of each other and to rebound without energy loss from the walls of the containing vessel.

In any closed vessel such as that shown in Figure 3.2, the motion of the molecules at any instant can be shown to be roughly equivalent to two equal and opposite streams, moving in each of three mutually-perpendicular directions. One of these

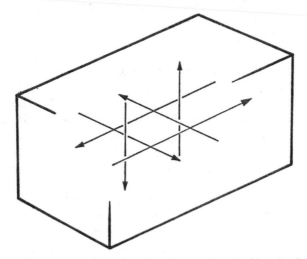

Fig. 3.2 Representation of molecular motion in kinetic theory

pairs of streams is considered to be moving as shown, parallel to the side of the vessel, whose length is l. Each molecule is assumed to have a mass m and average velocity v. If there are N molecules in the vessel, and thus $N/6$ in one stream, the number which strike the end in a brief interval of time t will be the fraction of those contained within a length vt of the vessel; that is $Nvt/6l$. The pressure in a closed vessel does not continually fall, so the molecules must not lose energy as they rebound. Thus, the velocity v is simply reversed in direction during inter-action with the wall. The MOMENTUM of the molecule is the product mv. This will have changed from mv towards the wall to mv away from the wall; a change of $2mv$ for each molecule. The total change of momentum of all the molecules in the time t is thus $Nvt/6l \times 2mv$. According to Newton's second law of motion, the reaction on the end is equal to the *rate* of change of momentum of the molecules, that is, the change per unit time, which is $2Nmv^2/6l$. The pressure p is the intensity of the reaction per unit area, that is

$$p = 2Nmv^2/6Al. \qquad\qquad 3\cdot1$$

But Al is the volume V of the vessel, so that

$$p = \tfrac{1}{3}\frac{N}{V}\, mv^2. \qquad\qquad 3\cdot2$$

The value is evidently the same for the other walls of the vessel. Since the mean kinetic energy E of a molecule is $\tfrac{1}{2}mv^2$, we have finally that

$$p = \frac{2NE}{3V}$$

or $\qquad\qquad\qquad\qquad\qquad\qquad\qquad\qquad\qquad\qquad 3\cdot3$

$$pV = \frac{2}{3}NE.$$

Now the pressure p and volume V of a body of gas are easily measureable quantities. As long ago as 1660 BOYLE had noted that, if a quantity of gas is maintained at a constant temperature (regardless of how temperature is defined) the product pV is found to be constant. If the gas if then made hotter or colder, pV increases or decreases. Since, from equation 3.3, pV has been shown to be proportional to the mean kinetic energy of a molecule, it is a short step to associate temperature with this quantity. Hence, we come to a definition of temperature in terms of the properties of an ideal gas. This is, that the temperature shall be taken to be directly proportional to the mean kinetic energy of its molecules. Then we could write

$$E = \text{a constant} \times T. \qquad\qquad 3.4$$

This is the simplest kind of relation between E and T which has the required properties, but as equation 3.4 is merely a definition of T, we could have chosen any other convenient relation. Using equation 3.4, we can then write equation 3.3 in the form

$$pV = \text{a constant} \times NT,$$

or $\qquad\qquad\qquad pV = kNT. \qquad\qquad 3.5$

The magnitude of the constant k depends upon the units used in measuring p, V and T. It is one of the fundamental constants of physical science and is called BOLTZMANN'S CONSTANT after the Austrian physicist of that name (1844–1906).

It will be seen that the definition of k implies that the mean kinetic energy of a molecule of an ideal gas has the value $3kT/2$. Examination of the methods of kinetic theory will now show that we would come to the same result whatever the nature of the particles of which the gas consists. We have spoken of them as molecules here, but they could have been single atoms or even elementary particles like electrons; whatever their nature, they have the same energy at the same temperature. They will, however, differ from each other by mass. A measure of relative mass among molecules is the quantity called MOLECULAR WEIGHT, M_1. (This is the mass of a molecule relative to that of the standard, which has been chosen to be the carbon atom whose weight is given the value 12). Then

$$m = \text{a constant} \times M_1 = k_1 M_1, \text{ say.} \qquad 3.6$$

The total mass M of gas in the vessel is evidently Nm, or $k_1 N M_1$. Thus the number of molecules present is $M/k_1 M_1$ so that equation 3.3 could be written

$$pV = \frac{M}{M_1} \frac{k}{k_1} T$$

which can be cast into the form

$$pV = M \frac{R_1}{M_1} T \qquad 3.7$$

by combining the constants k and k_1 into the single constant R_1, which is called the UNIVERSAL GAS CONSTANT.

Equation 3.7 is called the *equation of state* of an ideal gas. Such a relationship allows us to assign numerical values to temperature. For a fixed mass of gas in a closed vessel of constant volume, we would have, from equation 3.7, that the temperature is proportional to the pressure. That is, if the temperature is given the value of T_1 at some pressure p_1, the temperature would have the value T when the pressure is p such that

$$T = \left(\frac{T_1}{p_1}\right) p. \qquad 3.8$$

We might note that this relationship is the law obtained experimentally by CHARLES (1787).

For the Kelvin temperature scale, the value 273·16 K is

assigned to the temperature at which occurs a reproducible physical state known as the TRIPLE POINT OF WATER. This can be used to calibrate any gas thermometer so that its temperature is then given by equation 3.8, using only measurements of pressure which are easily made. The constant (T_1/p_1) may be chosen arbitrarily but the value 273·16 per atmosphere was used so that there continue to be 100 degrees on this scale between the freezing and boiling points of water at atmospheric pressure, as in the former centigrade or Celsius scale. Throughout this book we shall suppose that temperatures are measured on the Kelvin scale, and could be determined in principle by allowing an ideal-gas thermometer to come into thermal equilibrium with the system concerned. This scale is named after William Thompson, later LORD KELVIN (1824–1907).

3.7 Specific heat

In the next few chapters, we shall have occasion to use the term SPECIFIC HEAT. The reader will have encountered it before, but we ought to examine it because it has certain restrictions and qualifications even within the fairly narrow scope of our present needs.

The specific heat is the amount of energy required to increase the temperature of unit mass of a substance by one unit of temperature. Suppose that an amount of energy Q passes into our closed vessel containing an ideal gas. When conditions have become uniform, the energy will have been shared out among the N molecules present, and thus stored as an increase in the internal energy, denoted by U. That is, we could write

$$Q = \Delta U \qquad 3.9$$

(where the symbol Δ means 'the change of').

In the case of an ideal gas, the only form of internal energy is the kinetic energy of the molecules, so that

$$\Delta U = N \, \Delta E, \qquad 3.10$$

and since the total mass M of the gas is proportional to N and ΔE corresponds to a change in temperature ΔT, equation 3.9 can be reduced to

$$Q = \text{constant} \times M \, \Delta T,$$

which we could write in the form

$$Q = c_v M \, \varDelta T. \qquad\qquad 3.11$$

The constant c_v is called the SPECIFIC HEAT AT CONSTANT VOLUME. The qualification 'at constant volume' is necessary because if we heat a gas in a vessel whose volume can vary, the change in temperature will be different. For example, if the gas were in a balloon, the increase in pressure would cause an expansion of the balloon with an increase in the tension in the fabric. Then some of the energy has gone into straining the fabric and less is available for increasing the internal energy of the gas. For such a case, the specific heat would be greater than for the constant volume case.

Equation 3.11 can be used to define the specific heat of any substance. When a solid is heated, there are few cases in which its expansion is totally restrained and in most cases it is entirely unrestrained. Then the specific heat, whose value would be appropriate to such cases, would be the SPECIFIC HEAT AT CONSTANT PRESSURE. Evidently we can have other specific heats with other qualifications.

3.8 The ideal gas and other substances

We have developed our ideas in terms of the ideal gas because of its simplicity. It is possible, through the kinetic theory, to predict its behaviour from first principles. Although the behaviour of real elementary gases closely follows that of the ideal gas, every increase in complexity makes the kinetic theory less appropriate. Even in quite simple compound gases, such as carbon dioxide, CO_2, the molecule can store energy in other ways than by kinetic energy of translation. The molecule is big enough to have significant kinetic energy of rotation and can store energy in vibrational motions of several kinds. By using part of the argument of section 3.5, we can show that the constant c_v in equation 3.11 would be expected to have the value $3R_1/2M_1$ for an ideal gas. A development of kinetic theory predicts that for a molecule consisting of three atoms, c_v would have the value $3R_1/M_1$; the higher value arises from its additional freedom to store energy by rotation. However, we find that even a simple real gas like CO_2 has a value about 15% greater than this at ordinary temperatures. For any liquid or solid, we should not expect to be able to calculate the specific heats accurately from first principles without using a much more elaborate model, because of the multitude of ways in which energy can be stored in these systems.

Kinetic theory and the ideal gas are, nevertheless, of immense importance, because they have helped us to develop, in terms of prior concepts with which we are already familiar, new ones such as temperature and specific heat, which we can apply to other, more real but complex systems. Elsewhere in this book, when we come to develop other thermodynamic concepts, we shall begin again by reference to the ideal gas for which the consequences of a new step can be readily worked out.

3.9 *Units of thermodynamic quantities*

We are now armed with an account of most of the thermodynamic quantities used in the following chapters. Others will be discussed as they arise. When we wish to give numerical magnitude to these quantities, standard units must be used.

For the fundamental quantities of MASS, LENGTH, TIME and TEMPERATURE, we shall use the units and symbols of the Système Internationale d'Unités, which are respectively, the KILOGRAMME, kg, the METRE, m, the SECOND, s, (but sometimes the HOUR, h) and the KELVIN, K. When we have to specify a temperature in kelvins, it will be written TK; conversion to the value on the centigrade or Celsius scale, with the freezing point of water as the origin, is given by

$$T°C = TK - 273. \qquad 3.12$$

An interval of temperature of magnitude t on the Kelvin scale will be written t kelvins or tK. It should be recalled that it has been arranged for one kelvin to be of the same magnitude as a degree Celsius so that $1K = 1$ deg C.

The standard unit of ENERGY (and hence of HEAT and WORK) in the Système Internationale is the JOULE, J, and that of POWER, or rate of change of energy with time, is the WATT, W, equivalent to one joule per second. The joule is a rather unfamiliar unit at the present time, as well as being inconveniently small for our purposes—the energy required to heat one's home throughout a winter's day, for example, would amount to some thousands of millions of joules. We are all familiar, however, with the watt as a unit of power, our lights and electrical appliances being invariably rated in watts. Moreover, we are charged for our electricity consumption by the KILOWATT-HOUR, kWh, the familiar 'unit' of the electricity bill. In this book, then, we are using as our unit of energy the kWh, which represents the energy passing through a one-kilowatt electric fire in one hour. Where necessary, this can be reduced to the standard unit through the relation 1 kWh = 3·6 million joules.

In Chapters 1 and 2 we spoke of a world energy demand in hundreds of millions of millions of kWh. Though this is a perfectly adequate way of writing large numbers, it is rather a clumsy one. The Système Internationale includes a series of prefixes to simplify the presentation of such numbers—mega (M) for million, giga (G) for thousand million, tera (T) for million million and so on. A similar series is used for small numbers—we have already encountered the prefix for one millionth, for example, in the micrometre (μ). Although these prefixes are already part of the everyday language of science, they are not very familiar in the world at large and apart from a few (micro, milli, kilo, mega, perhaps) they will remain obscure when they are not used constantly. It would not be meaningful to many, for instance, if our estimate of the probable future energy demand were given as 200 T kWh. Later, we shall meet the same difficulty with very small numbers. The charge on the electron is about 0·16 attocoulomb, and we would have to render this as about one-sixth of a thousand million millionth of a coulomb. One way out of this dilemma is to use the powers-of-ten notation ($10^2 = 100$, $10^4 = 10\,000$, $10^7 = 10\,000\,000$ and so on). This notation is brief, universally recognised and demands no recollection of prefixes. Thus, our energy demand figure could be written 2×10^{14} kWh (or 2×10^{17} Wh) and the charge on the electron as $1·6 \times 10^{-19}$ C, 10^{14} meaning (one followed by 14 noughts) and 10^{-19} meaning the fraction one divided by (one followed by 19 noughts).

In the rest of this book, then, we shall continue to use the familiar terms thousand and million in most cases, but where this threatens to become too clumsy, we shall revert to the powers-of-ten notation. This will serve us wherever our study leads, from the unimaginably small to the inconceivably large.

4

COLLECTION OF SOLAR ENERGY

So quick bright things come to confusion.

WILLIAM SHAKESPEARE (1564–1616)

IT is a matter of common experience than an object standing in sunlight becomes warm. There are many uses to which this occurrence could be put, either directly—as in heating a house or cooking food—or indirectly, as in the generation of electricity. We shall want to enquire whether such processes are economically feasible. To begin with, we need ways of calculating the increase in temperature in particular cases. Before we can do this, we must look in some detail at the manner in which radiation interacts with material bodies on which it falls.

An object standing in the sun warms up because it absorbs some of the energy transported by the solar radiation. The mechanisms by which it can do this are many, and a satisfactory description of them was not possible until this century, with the introduction of the QUANTUM THEORY. Though this theory, and that of WAVE MECHANICS which grew out of it, has advanced enormously in complexity, we can still obtain a working understanding of it in terms of quite simple models, such as that of Bohr for the atom. Nothing more elaborate is needed here.

4.1 *Elements of quantum theory*

Quantum theory is bound up with the wave-particle duality exhibited by things on a very small scale. We have seen in a previous chapter that in many of its activities, light displays its wave-like character and its wavelength λ is easily measured. NEWTON (1704), however, first though of light as a stream of particles or corpuscles, and this view is revived in the notion of the photon, conceived by EINSTEIN (1905) to explain how light could cause electrons to be knocked out of atoms. In this PHOTO-ELECTRIC phenomenon, light shows properties of particles, whose energy depends upon the wavelength exhibited by the light in its role as a wave. All other electromagnetic radiation has similar properties. The complementary notion that objects which normally seem to behave like particles would have a corresponding wavelength, dependent upon their energy, is due to DE BROGLIE (1924). It is now known that all very small things have this dual character, even things as relatively large as whole atoms sometimes displaying wavelike properties.

The essential point of quantum theory is the assumption that the energy of any system showing wave-like characteristics can only change in multiples of a certain definite amount. This minimum change, called a QUANTUM, is related to the FREQUENCY of the associated wave characteristics. In fact, it is found to be directly proportional to it. Now when a wave, of wavelength λ, is travelling with velocity v, the frequency, which is the number of waves passing a given point in unit time, is evidently

$$\nu = v/\lambda. \qquad 4.1$$

For example, the velocity of light waves is about 3×10^8 m/s and the visible wavelengths occur around 0.6×10^{-6} m so that the frequencies are around 5×10^{14} cycles per second.

The magnitude of the quantum of energy for such a wave is given by

$$E = h\nu, \qquad 4.2$$

where h is a constant known as PLANCK'S CONSTANT (after MAX PLANCK (1901), who had obtained a relation similar to 4.2 for another purpose.) This constant is exceedingly small, 6.6×10^{-34} J-s, so that only for very high frequencies is the quantum of energy large enough for discontinuous changes to be noticeable. As we can see, the quantum for light, which is the energy of one photon, is about 3×10^{-19} J. Only on an atomic scale do we find energy changes of this order to be important.

So far we have been concerned with the treatment of energy in quantum theory. The next property to be considered is MOMENTUM, the product mv of the mass of a particle and its velocity. When we dealt in Chapter 3 with the kinetic theory of gases, we obtained a relation between the momentum of the gas molecules and the pressure on the walls of the containing vessel. Sensitive apparatus is able to detect a pressure when light falls on an object, and this, too, can be associated with the momentum of the light photons. From the way that the pressure of light varies with the wavelength λ, it is found that the momentum q of a radiation photon is related to its associated wavelength by the expression

$$q = h/\lambda, \qquad\qquad 4.3$$

where, as before, h is Planck's constant.

When de Broglie came to review the state of quantum theory in 1924, he proposed that the wavelength of the wave associated with any particle would also be related to the momentum in the same way as for radiation, i.e., according to equation 4.3. This has been amply confirmed since that time by numerous experiments.

Our object in this Section is to prepare for a consideration of the manner in which radiation interacts with material bodies. Mention has already been made of the photo-electric effect, in which light is sometimes able to knock electrons out of atoms. To understand this and similar phenomena, we may turn to the simplest model of the atom which will serve our purpose, the Bohr model. In this, we visualise the electrons as classical particles retained in orbit round the nucleus by the electrostatic force known as the Coulomb force, arising from their negative charge and the positive charge of the nucleus. Any motion of a particle in a circular path requires an inwardly-directed force such as this, to maintain it. A particle moving freely, without being influenced by any forces, travels in a straight line—as stated by Newton in his first law of motion. When a particle is being made to travel in a circular path, it is easily shown that it has at all times an acceleration, which has the value v^2/r directed towards the centre of the path, when the velocity is v and the radius of the path r. Then according to Newton's second law, a force, proportional to the mass and the acceleration, must be acting on the particle in an inward direction to make it continue this kind of motion. The electrostatic force on the electron is proportional to the charge Z on the nucleus and varies inversely as the square of the distance from it, so that for a steady orbit we shall require, to satisfy the second law, that

$$CZ/r^2 = mv^2/r, \qquad\qquad 4.4$$

where C is a constant representing the proportionality of the Coulomb force to Z.

Thus the radius of the path is given by

$$r = CZ/mv^2. \qquad 4.5$$

Now the momentum of the electron is mv and its de Broglie wavelength, by equation 4.3, is, therefore

$$\lambda = h/mv. \qquad 4.6$$

In this model, if the behaviour is not to vary with time, there must be a whole number of these wavelengths in the circumference of the orbit. If this number is n (n = 1,2,3,4 etc.) we must have

$$2\pi r = n\lambda, \qquad 4.7$$

and combining this with equations 4.5 and 4.6 to eliminate λ and v, we find that

$$r = \frac{n^2 h^2}{4\pi^2 C Z m}. \qquad 4.8$$

Thus, only certain orbits are available to the electron, with radii given by equation 4.8. for each integral (whole number) value of n. Now each electron has kinetic energy because of its motion, and potential energy by virtue of the electrostatic attraction between it and the nucleus (since work would have to be done on it to increase its distance from the nucleus). By developing this theme, it can be shown that the *total* energy E of the electron is given by an expression having the form

$$E = A - (CZ/2r), \qquad 4.9$$

where A is a constant depending upon the (arbitrary) position at which the potential energy is said to be zero. Then from equation 4.8, the energy is

$$E = A - \frac{2\pi^2 C^2 Z^2 m}{n^2 h^2}. \qquad 4.10$$

As each successive value of n (1, 2, 3 etc.) is inserted we obtain a series of values of energy in an increasing progression. These are known as the PERMITTED ENERGY LEVELS of the electron. When it has one of these values of energy, the electron is said to be in a STATIONARY STATE; as long as it remains in the atom at all, it must be in one of these states.

When such an electron is moved from one orbit, say at radius r_1, to another, say at radius r_2, we can see from equations 4.9 and 4.10 that its energy increases by an amount

$$\Delta E = \frac{CZ}{2} \left(\frac{1}{r_1} - \frac{1}{r_2} \right)$$

$$= \frac{2\pi^2 C^2 Z^2 m}{h^2} \left(\frac{1}{n_1^2} - \frac{1}{n_2^2} \right). \qquad 4.11$$

If it has acquired this energy by absorption of radiation, it must have been by capturing a photon whose energy is just ΔE. From equation 4.2., the frequency of the associated wave of this photon must have been

$$\nu = \frac{\Delta E}{h} = \frac{2\pi^2 C^2 Z^2 m}{h^3} \left(\frac{1}{n_1^2} - \frac{1}{n_2^2} \right). \qquad 4.12$$

Beginning at any orbit whose PRINCIPAL QUANTUM NUMBER n_1 is given, we may then determine the spectrum of frequencies (and hence the wavelengths) of the radiation which is capable of moving an electron to a series of further orbits, specified by n_2. Evidently, if the energy of the incoming photon is greater than that required to reach the outer orbit, the electron will be ejected altogether, as in photo-electric behaviour.

We have been using a very simple model of the atom here, as a means of introducing the ideas of energy levels, photon absorption, etc. In fact, the hydrogen atom, which has only one electron, is very well represented by Bohr's model and its behaviour is exactly as predicted. For more elaborate atoms, the model is too simple. We may, nevertheless, begin to generalise a little, after this introduction, to the behaviour of more complicated systems, bearing the crudeness of our assumptions in mind. In more modern developments of quantum mechanics, it has been possible to abandon the concept of electrons as particles orbiting the nucleus. But the notion of ENERGY LEVELS of the atom has been carried over from the simple model and is an essential element of more recent work.

4.2 Interaction of radiation with bodies

We have seen that an interaction can only occur between an incident photon and an electron if the photon has the correct energy to move the electron to another acceptable orbit. Then we find that the atom shows LINE ABSORPTION; on a spectrum of wavelength, absorption occurs only at particular wavelengths satisfying the quantum rules. In a solid, however, there are many more energy levels available than in an isolated atom. The atoms are sufficiently close together for the electrons to be influenced by other nuclei. In electrical conductors in fact, the outer or valence electrons are able to drift about as if they belonged to no particular atom. Moreover, the atoms themselves have freedom to vibrate, which adds yet more energy levels to the possible states. As a result, most solids do not show line absorption, the wavelengths of photons which can be absorbed being so numerous that they merge into a virtually continuous spectrum.

Not all of the photons incident on a solid will be absorbed, however. Some will be scattered in near-collisions and others might pass right through the body.

The likelihood of absorption of a photon varies with its energy, and hence its wavelength, even for solids with an effectively continuous absorption spectrum. When the number passing through, either directly or after scattering, is a substantial fraction of those falling on the surface, the material is said to be DIATHERMANOUS (literally 'capable of passing heat', and therefore an unfortunate choice of term). In more common usage, when we speak of visible radiation only, we call a substance TRANSPARENT if most of the incident photons pass through directly or TRANSLUCENT if many are scattered through without absorption.

Some photons are scattered back and re-emerge again from the surface upon which they were incident. If the material is diathermanous, some of these photons will have been scattered about inside the body several times, but if it absorbs strongly, only those photons scattered backwards near the surface will reappear. In both cases, we say that the emerging photons have been *reflected* by the body. Thus the fate of incident photons will be that they are either reflected, absorbed or transmitted. The fractions of the incident photons found in each category are called the REFLECTIVITY, ABSORPTIVITY and TRANSMISSIVITY of the body respectively.* It follows that these

* These are the terms most commonly used in radiometry in the U.K. and will be used here, though they have now been superseded by the terms

68

fractions must add up to unity. They all depend upon the wavelength of the radiation, as we have seen, but also upon the direction of the radiation and the condition of the surface of the body. If the surface is very smooth, for example, the reflectivity can be high, particularly for radiation approaching at glancing incidence, but if it is rough, photons may be trapped in the cavities of the surface and have more chance of being absorbed before being able to re-emerge.

4.3 *Emission of radiation*

When a body absorbs radiation, it is raised to an excited state, with electrons at high energy levels and lattice vibrations proceeding vigorously—that is, its temperature is raised. The incident photons—'quick bright things' —have vanished in the confusion. The body seeks to restore its original state by re-radiating this extra energy. The process now takes place in reverse; the photons emitted again have particular wavelengths according to the energy change, but the number of possible values is often so great that we may take the emission spectrum to be continuous. When an atom or molecule is raised to an excited state, it usually remains in it only for the briefest instant of time, after which it is likely to re-emit the photon. However, in a solid body or a dense gas, it may, before it can do this, transmit the energy to neighbouring atoms through the action of the interatomic forces. By this means, the temperature is evened out and the excitement of the body becomes more uniform. As a result of the re-distribution of the energy, the emitted radiation may have a different wavelength distribution from that of the absorbed radiation. In particular, the distribution is now controlled principally by what we call the temperature of the body as a whole.

The interaction between radiation and any given body is so complex that its prediction from first principles is quite impracticable. It is therefore convenient to consider, instead, the behaviour of an ideal standard, or model, body having specified and rather simple properties. If such a body has an infinite number of permitted energy levels, it is termed a BLACK BODY,

---

REFLECTION FACTOR, ABSORPTION FACTOR, and TRANSMISSION FACTOR for international usage. American writers have often used as alternatives the terms REFLECTANCE, ABSORPTANCE and EMITTANCE.

because it is able to absorb all radiation falling upon it, whatever the wavelength. The prediction of the radiative properties of the black body of MAX PLANCK (1901) was the first instance of the use of quantum ideas, and was one of the most outstanding conceptual steps in the history of physics. We shall not attempt to reproduce the argument here, but merely state that it shows that the radiation emitted by a black body, by virtue of its temperature, has a particular distribution of energy density D—as defined for the sun's radiation on page 26—with respect to wavelength. We can express this in the form*

$$D = T^5 \times \text{some function of the product } (\lambda T) \qquad 4.13$$

where λ is the particular wavelength concerned and T is the absolute temperature of the body. It is to be noted that the energy density depends upon the product (λT) in an important way.

The function of (λT) in equation 4.13 is illustrated in Figure 4.1. and this shows how the radiation varies with wavelength for any given temperature. (We should recall here that it is the absolute temperature which is important; that is, the temperature measured on a scale which begins at the point at which no radiation is emitted, the absolute zero.) The maximum energy density occurs near the point $(\lambda T) = 2900$ μm K. It is plain then, that the wavelength λ at which the maximum occurs decreases with increasing temperature—it is $2900/T$ K μm. In fact, the whole curve is displaced with respect to wavelength as the temperature changes. For example, if the sun were a true black body radiator, with a surface temperature of 5800 K, its maximum energy density would be found at $\frac{1}{2}$ μm (as it is, in fact) but for a black body at a typical earth temperature of 290 K (17°C), the maximum is at 10μm, far into the infra-red. We first notice that a body is hot by its radiation alone when the energy density near the red end of the visible spectrum is high

* The actual relation is

$$D = c_1 T^5 / (\lambda T)^5 \{ \exp (c_2/\lambda T) - 1 \}.$$

If D is in watts per square metre per μm, T in kelvins and λ in micrometres, the constants c_1 and c_2 have the numerical values 3.74×10^8 and 1.44×10^4 respectively. exp x is the exponential function e^x, where e has the value 2.7183.

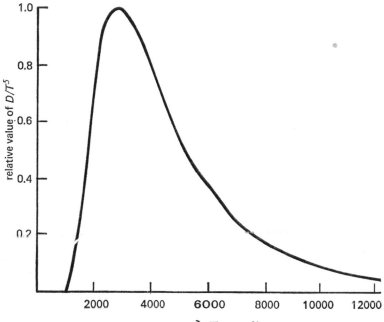

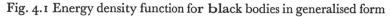

Fig. 4.1 Energy density function for black bodies in generalised form

enough for it to be bright in relation to the surroundings. This occurs at temperatures around 1500 K (about 1200°C) when the energy-density maximum comes at about 2 μm. We then say that the body is red-hot. We can, of course, *feel* the radiation at much lower temperatures through its warming action on the skin. In a room at ordinary temperatures, it is possible to feel thermal radiation from an object only some 10 K hotter than the surroundings.

The total rate of energy emission by a black body can be obtained by adding together the contributions at all wavelengths whose values are given by the curve of Figure 4.1. The actual rate of emission at any wavelength also involves T^5, as shown in equation 4.13, but when the whole is worked out, it is found that the rate of emission of energy per unit area, or the INTENSITY, is proportional to the fourth power of the absolute temperature. Then we have symbolically

$$P_e = \sigma\, T^4, \qquad\qquad 4.14$$

where the constant of proportionality, σ, is called the STEFAN-BOLTZMANN CONSTANT.

A useful form of this equation is

$$P_e = (T/64\cdot5)^4 \text{ watts per m}^2, \qquad 4.15$$

when T is in kelvins. (That is to say, σ has the value $(1/64\cdot5)^4$ watts $/\text{m}^2 \text{ K}^4$). Since P is a *rate* of passage of energy, we note that it represents *power* and is given in an appropriate power unit such as the watt.

For real bodies, with a complicated distribution of energy levels, we usually find that the radiation is not distributed quite like that of a black-body, either with respect to wavelength or to direction of emission. However, for simplicity, we sometimes use the black body as a standard and represent the gross properties of a body in terms of this. Thus we might assign to the body an overall EMISSIVITY, ϵ, such that at a temperature T, it emits a fraction ϵ of the energy emitted by a black body at that temperature. Further, we assign the properties REFLECTIVITY, ρ, ABSORPTIVITY, a, and TRANSMISSIVITY, τ, to a body such that if radiation of intensity P falls upon it, the rates at which energy is reflected, absorbed and transmitted are respectively ρP, $a P$ and τP. We note that all the properties ϵ, ρ, a and τ lie between 0 and 1 for real bodies, though for a true black body they would have the values 1,0,1 and 0 respectively.

These radiative properties vary widely between bodies and, most importantly, vary with the wavelength of the radiation for a given body. This wavelength dependence is, of course, due to the character of the absorption-emission process. It is conveniently represented by giving properties in terms of the temperature of the body (for emissivity) or the temperature of the source of the incident radiation, and thus its wavelength (for other properties). Some typical values are given below in Table 4.1.

It is found, in general, that polished metals have low emissivity at all temperatures, though their behaviour is greatly changed by surface treatments, presence of oxide films etc. Paints, which are clearly distinguishable to the eye by colour, can have high or low absorptivity for radiation similar to that of the light coming from the sun. But when exposed to long-wave radiation, or when at low temperatures, their absorptivity and emissivity are uniformly high except for those pigmented with

TABLE 4.1

Approximate values of radiative properties of various substances

| Substance | Temperature of body or of radiation source | | | | |
	20–100°C			5 000°C	
	ρ	α	ϵ	ρ	α
Polished metals	0·9	0·1	0·1	0·7	0·3
Oxidised metals	0·2	0·8	0·8	0·8	0·2
White gloss paints	0·1	0·9	0·9	0·8	0·2
Black matt paints	0·05	0·95	0·95	0·1	0·9
Aluminium paint	0·5	0·5	0·5	0·8	0·2
Concrete	0·1	0·9	0·9	0·4	0·6
Tiled roof	0·1	0·9	0·9	0·2	0·8
Glass	0·1	0·9	0·9	0·1	0

aluminium or other metallic flakes. For this reason, the efficency of a domestic radiator is not much affected by the colour of the paint used to decorate it, since it operates at low temperatures and emits long-wave radiation. The radiative properties of glass are extremely important. We have exploited glass because it is virtually transparent to the short-wave radiation we see by. At longer wavelengths, however, it is almost opaque and we shall see later that we can greatly profit by these characteristics.

4.4 *Equilibrium temperature of irradiated bodies*

We can now return to the behaviour of a body left lying in the sun. If we consider the ways in which it can gain or lose energy, shown in Figure 4.2, we can see that the situation is quite complicated. So that an understanding of the behaviour can be obtained quickly, we shall at first simplify the situation a little. Suppose that the body in question were a thin plate lying on an insulated base, as shown in Figure 4.3(a). We have here the elements of what is called a FLAT-PLATE SOLAR COLLECTOR. It receives energy from the sun and radiates it away again. We have arranged it so that it cannot "see" the surroundings and so it exchanges no radiation with them. For the moment, we will neglect heat exchange with the atmosphere by convection and radiation, returning to them later.

If the solar radiation is of intensity P and the plate has an

73

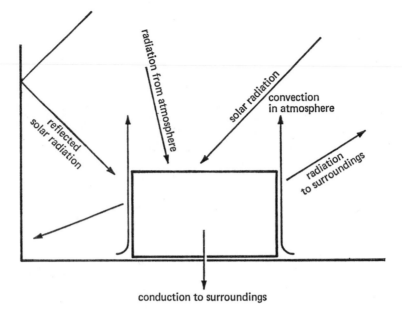

conduction to surroundings

Fig. 4.2 Energy flows for irradiated body

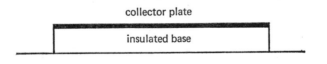

a)   simple flat-plate collector

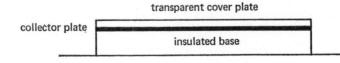

b)   collector with cover plate

Fig. 4.3 Flat-plate solar collectors

74

absorptivity α_s for this radiation, the plate will heat up until it reaches an equilibrium temperature T. At this temperature, the rate of emission of radiation just balances the input, and we shall have the equality

$$\alpha_s P = \epsilon \sigma \, T^4 \, , \qquad\qquad 4.16$$

where ϵ is the emissivity of the plate at low temperatures. Then the equilibrium temperature T is given by

$$T^4 = \frac{\alpha_s}{\epsilon} \, \frac{P}{\sigma} \, . \qquad\qquad 4.17$$

To obtain the highest temperature, we evidently need a high ratio of α_s/ϵ. From the data of Table 4.1, we see that, in the case of polished metals, this ratio can sometimes be as great as 2 or 3, but often this is not the only consideration. If the device is to be used for collection of solar energy, for instance, we shall also need a fairly high value of absorptivity, which we do not get with polished metals. For surfaces with high absorptivity, it is usually found that α_s/ϵ is about unity. For reasons which will appear later, we shall call such a surface a NEUTRAL ABSORBER. Then taking this value and $P = 800$ watts/m², a typical summer value for a tropical region, we obtain from equation 4.17 that the equilibrium temperature is 343 K (70°C). In spite of the simplifications here, this is a fair estimate of the temperature reached by a black plate left for a time in the tropical sun.

We can now refine the method a little. One factor of importance is the heat loss to the air by convection. The basic mechanism of this is simple. Air in contact with the hot surface becomes warmed and expands; its density is thus reduced relative to that of the surrounding air and it rises because it is buoyant. Cooler, denser air moves in to take its place and a convection current is established. (This kind of process is often rendered visible over a heated surface, like a road, by the refraction of light produced by changes in density of the air). In still air, the rate of heat loss by this mechanism, from a small surface, horizontal or slightly inclined, is about 4 watts/m² for each K difference between the surface and the undisturbed air, though it is not exactly proportional to the temperature difference, as implied by this. When the air is driven over the surface—if a wind is blowing for instance—the rate of heat loss is greatly increased. At a wind speed of 10 m/s (about 20 mph) for example, the rate of loss becomes about 30 watts/m² per K for small temperature differences. For the moment we might regard these as values measured by experiment. Though heat loss by convection can be satisfactorily described analytically, the theory is rather complicated and would be an unnecessary diversion here.

Another factor, which works in the opposite direction, is the presence of long-wave radiation from the atmosphere. This is the re-emission, principally by molecules of carbon dioxide and water vapour, of the energy absorbed by these components from the incoming solar radiation, and by radiation and convection from the earth. These molecules have vibrational and rotational energies for which the quantum changes correspond to the energy of certain photons in the visible and infra-red wavelengths. Some of the energy is re-distributed before re-emission, but the emission spectrum for the atmosphere still shows fairly distinct lines and bands corresponding to the absorption wavelengths.

The total intensity P_a of this radiation depends rather strongly on the amount of water vapour present, particularly near the ground, and is therefore very variable. In very damp conditions with complete cloud cover, the atmosphere behaves almost like a black body with a temperature around 280 K (10°C); the corresponding radiation intensity on a horizontal surface is about 300 W/m². In the United Kingdom, the long-wave radiation from the atmosphere reaches this value on occasions. In very clear conditions with an arid climate, the radiation is more irregularly distributed with respect to wavelength, being concentrated round the CO_2 absorption band at 15 μm and the H_2O bands lying between 20 μm and 100 μm. Even so, the total intensity is rarely as low as 100 W/m². This very substantial input of radiation to exposed surfaces is responsible for the maintenance of reasonable temperatures during the hours of darkness. Without it, the surface temperature would fall rapidly by radiation to space, as it is observed to do on the moon, which has no atsmophere. On the earth, the surface temperature may fall enough during the night to produce a RADIATION FROST, but rarely falls by this means much below 0°C except towards the poles.

We can now see why the first estimate of equilibrium tempera-

1. Largest solar furnace in the world, Odeillo, France.

2. 36m² solar heater serving 120m³ swimming pool, Kent, England
(*designed by Edward J. W. Curtis, RIBA, Dip Arch*)

3. 2.2m² solar heater for domestic hot water supply
(*designed by Edward J. W. Curtis, RIBA, Dip Arch*)

4. 2.5m³/h solar still at the Bakharden State Farm, Turkmenian SSR

5. Inflatable solar still for aircrew

6. Photoelectric cell array for spacecraft, Earth Resources Technology Satellite, ERTS1

7. Stone-Chance solar powered navigation aid, Arabian Gulf

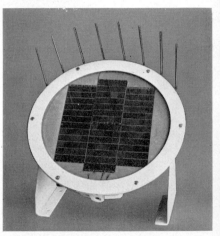

8. Photo-electric cell for Stone-Chance navigation aid

9. Solar powered boat for remote areas

ture by equation 4.17 was reasonable, although we had not included heat exchange with the atmosphere. It gave a temperature of 70°C. If the ambient temperature were 17°C and a convection loss occurred at a rate of 4 W/m² °C, this loss is 212 W/m², which is just about balanced by the long-wave radiation gain in typical circumstances.

4.5 *Improvements to the flat-plate collector*

As we saw earlier, the performance of the flat plate collector, determined for the moment solely by its equilibrium temperature, could be improved if the ratio of absorptivity of solar radiation to emissivity of long-wave radiation could be increased, without reducing the absorptivity too far. This can be achieved by producing what are called SELECTIVE ABSORBERS. These are usually polished metal surfaces coated with thin deposits of black salts such as the oxides of nickel and copper. These have a high absorptivity, of the order of 0·9, for short-wave radiation. If the layer is thin, however, it is found to be transparent to radiation of wavelength longer than the coating thickness. Then, for the long-wave radiation appropriate to absorber temperatures, the emissivity is close to that of the metal beneath, say about 0·1. With such a selective absorber, then, a_s/ϵ would be about nine, and the equilibrium temperature under the conditions assumed previously would rise to 427 K, 154°C (Taking the long-wave radiation from the atmosphere to be 200 W/m² and an absorptivity to this radiation of 0·1). This is a very substantial improvement, though not easy to achieve in practice. The chief difficulty is that many of the selective surfaces are sensitive to contamination by dust and do not keep their properties for long when exposed to the weather.

One of the most important further modifications made to the flat-plate solar collector is the fitting of one or more transparent cover plates above the absorber surface, as shown in Fig. 4.3b). The majority of glasses and clear plastics in thin sheets transmit about 90% of solar radiation, but only about 10% of radiation with wavelengths greater than about 2 μm. These characteristics give rise to the 'greenhouse effect', in which the glass allows the solar radiation to pass in, but absorbs the long-wave radiation emitted by the surfaces of the interior of the greenhouse. The convective heat loss to the air is also modified; there is now an intermediate stage involving convection within the air between the absorber plate and the cover plate. This should evidently be kept to a minimum. With a spacing of a few centimetres, the heat exchange by convection is found to be about 4 W/m² per K temperature difference between the plates in typical circumstances, and because the cover plate will be warmer than the surrounding air, this represents a useful saving. There is also an exchange between the plates by radiation, which, for the same reason, is more favourable than exchange with the atmosphere. As in the case of the convective heat loss to the surrounding air from the cover plate, the internal heat exchange by convection is not quite proportional to the temperature difference, though we shall assume it to be so here as a convenient approximation. The rate of 4 W/m² K which we have chosen, can be seen to be reasonable by comparing it with the rate of ordinary conduction through a layer of air 2 cm thick, which is about $1\frac{1}{4}$ W/m² K. The additional transfer of energy by convection currents between the absorber and cover-plates thus increases the rate by a factor of about 3 in this case.

To obtain estimates of the effects of these factors, we may legitimately make some further simplifications. Without too serious an error, we may suppose for the moment that the cover plate is totally transparent to solar radiation and totally opaque to long-wave radiation. Its emissivity will also be taken to be unity. It can be shown that the rate of exchange of energy between two plates, one a black body at temperature T, and the other with emissivity ϵ at temperature T_1, is given by $\epsilon\sigma(T^4 - T_1^4)$. We may then express the rate of exchange of energy P_1 between the absorber and the cover in the form

$$P_1 = h_1 (T - T_1) + \epsilon\sigma(T^4 - T_1^4), \qquad 4.18$$

where h_1 is the heat-transfer coefficient for the convective part of the exchange. Then the energy balance for the absorber plate is simply given by

$$a_s P = P_1 \qquad 4.19$$

and that of the cover plate by

$$P_1 + P_a = h(T_1 - T_a) + \sigma T_1^4, \qquad 4.20$$

where P_a is the intensity of long-wave radiation coming in from

78

the atmosphere. Then for any given conditions, we can obtain the equilibrium temperatures of the two plates: T_1 from equations 4.19 and 4.20 and then T from equations 4.18 and 4.19.

The effect of a single cover plate is considerable. For the conditions of our examples, with a neutral absorber and a selective absorber surface respectively, we find equilibrium temperatures of 386 K (113°C) and 467 K (194°C). Further improvements can sometimes be gained by the fitting of additional cover plates. As the number increases, however, the reduction in losses from the absorber plate is opposed by the reduction in the solar energy reaching the plate by absorption in the cover plates (which we ignored here for the case of a single plate). In practice, it is rarely considered worthwhile to use more than two cover plates for this reason.

Another small disadvantage of the cover plate is the reduction in transmission through it with increasing obliquity of the radiation. The reflection of rays at the surfaces of a sheet of glass may be predicted by using the laws of geometrical optics, the directions and intensities of the reflected and refracted rays being given by Snell's and Fresnel's laws respectively. For a thin sheet of glass, as used for windows, it is found that the transmission has a reasonably constant value of about 0·9 for rays incident at angles up to about 60° to the normal, and to a fair approximation the transmission may then be considered to decrease linearly with angle from 0·9 at 60° to zero at 90°.

A further increase in equilibrium temperature can be obtained by reflecting more of the solar radiation onto the absorber by mirrors. One of the simplest arrangements with plane mirrors is shown in Figure 4.4. The factor by which the intensity would be increased by perfectly-reflecting mirror systems is evidently given by the ratio of the total area of beam intercepted to that of the absorber plate onto which it is directed. This factor is called the CONCENTRATION RATIO. The mirrors should be mounted so that all the rays falling on them are directed on to the absorber plate, as shown. For a square absorber, fitted with four mirrors, each of the same size as the absorber plate (which permits folding when not in use) the angle β must be 60° and the concentration ratio is then 3. Not all of this benefit could be realised in practice because the reflectivity of the mirrors would not be 100% and because the cover plates, if fitted, reflect away more of the incoming radiation when it falls at a shallow angle on the surface. Nevertheless, a concentration ratio of 2 is readily obtained.

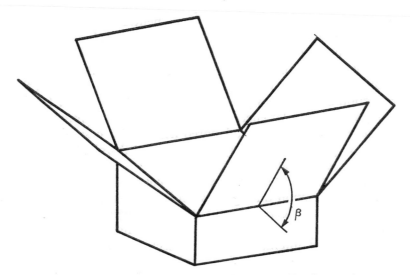

Fig. 4.4 Concentration of solar radiation with plane mirrors

Flat plate solar collectors with mirror systems giving a CR of 2, would, under the conditions of our running example, reach equilibrium temperatures of 180°C (neutral absorbers) and 322°C (selective absorber). We should note here that only the direct component of the solar radiation is enhanced by reflectors; the diffuse component cannot be redirected onto a given surface.

A final refinement to be considered is the use of a parabolic reflector, as shown in Figure 4.5. A perfect parabolic reflector, exposed to a parallel beam of radiation, concentrates it all at the focus. In such a case, the concentration ratio would be infinitely great. Now, whenever there is the appearance of being able to obtain something quite implausible—an infinite equilibrium temperature at the focus in this case—we must look for an imperfection which would prevent this in practice. The cause here is that the sun's rays are not exactly parallel. They appear to us to come from a disc whose width subtends an angle of about 32 minutes of arc. In natural angular measure, this is the ratio of the sun's diameter to its distance from us, or about

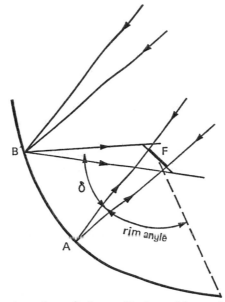

Fig. 4.5 Concentration of solar radiation with parabolic mirror

0·0093 radians; we call this angle γ. This is of no great importance for the flat mirror reflector, but it sets a limit on the concentration ratio of the parabola. The result is that the energy falls within a region around the focus rather than exactly at a focal point. We see this process in reverse where the finite size of filament in a vehicle headlamp results in a slightly diverging beam. In Figure 4.5 are shown the paths of the rays coming from opposite edges of the sun and impingeing at two points on the mirror, A and B. The rays reflected at the centre of the mirror A return along reciprocal paths, so that at the focus F, they produce an image of the sun of width $2f \tan \gamma/2$, which, since γ is a small angle, is about $f\gamma$. (i.e. for a focal length f of 2 metres, the image is about 2 cm wide). The rays reflected from other parts of the mirror, such as those passing through B, have a longer distance to travel before reaching F and have diverged more. Not only do they form a larger image, but this image is thrown on a plane inclined at the angle δ to the focal plane. On the focal plane itself, the image is elliptical. The total effect at the focus is a combinaton of a multiplicity of such ellipses, all bigger than the central region formed by the rays reflected at A. If the sun's disc were uniformly bright, the intensity would be uniform across this central region but would decrease rapidly with distance away from it. We can readily see that the contribution of rays reflected from a given area of other parts of the mirror becomes increasingly less as the angle δ increases. At the focal plane, these rays are more spread out and arrive at shallower angles to the plane. On the other hand, a given interval in δ encloses a greater area of mirror surface as δ increases. These factors work against each other, producing a variation of concentration ratio near the centre of the image, with angle, as shown in Figure 4.6. We will not attempt to

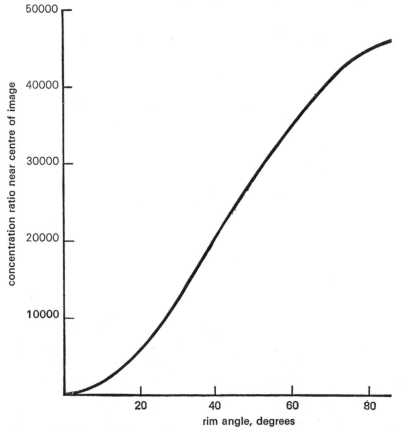

Fig. 4.6 Effect of rim angle on concentration raio near centre of image

calculate these values, but the previous discussion will lend them plausibility. We see that for mirrors with small rim angles, say less than 25°, the ratio grows slowly, and also that there is little further increase for angles greater than about 70°. Nevertheless, the concentration ratio for the central region near the focus is very large: about 10000 and 40000 respectively for the two angles quoted. (In practice, somewhat lower values are obtained, even for a perfect mirror, as a result of the variation in brightness across the sun's disc).

If a small body is placed at the focus of a parabolic mirror facing the sun, it will reach an equilibrium temperature which depends largely upon its radiation losses. Heat loss by convection is relatively unimportant. For a black body, insulated so that it radiates only from the exposed face, we would then have

$$P \times CR = \epsilon\sigma\, T^4. \qquad 4.21$$

Then if $P = 800$ W/m² as before, and with concentration ratios of 10000 and 40000, we find equilibrium temperatures of 3440 K (about 3170°C) and 4850 K (about 4480°C) respectively.

Because the radiative loss depends on T^4, the equilibrium temperature changes less with concentration ratio than might be expected. Even for $CR = 5000$, we find $T = 2890$ K (2620°C) with conditions as before. It is observed as a result that quite high temperatures can be obtained even with mirrors of indifferent quality. However, a high temperature may not be the only requirement.

If it is desired to secure the maximum amount of energy, the receiving body should be large enough to accept all the rays from the mirror. The outer rays, as we have seen, are of lower intensity than those at the centre. Moreover, the poorer the optical quality of the mirror, the larger the receiver must be and the lower the effective value of the CR and hence the equilibrium temperature.

In subsequent work referring to systems employing parabolic and other reflectors, it will be convenient to speak of the EFFECTIVE CONCENTRATION RATIO, without references to how it is to be realised in practice. We shall define this quantity so as to facilitate calculations of the kind given by equation 4.21. Thus, the power input to a surface at the focus will be $P \times CR$ per unit area. Values of the effective CR thus defined will depend upon the shape of the receiving surface and the fraction of the solar image covered. For mirrors of average quality, and receivers accepting most of the rays, the effective CR will not often be greater than 1 000 or so. The equilibrium temperature reached at CR = 1 000, according to equation 4.21, is about 1 930 K (1 660°C).

4.6 The solar pond

An ingenious variation on the flat-plate collector is the SOLAR POND, investigated by Dr. H. TABOR and his colleagues at the National Physical Laboratory of Israel. In this, the absorber is merely a pond of water, which can be fitted with a cover plate if desired. Upon exposure to solar radiation, the temperature of the water is raised by direct absorption of photons or by conduction from the absorbing tray which forms the bottom of the pond. As its temperature increases the water expands and, the hotter elements being lightest, they would normally rise to the surface. In this form, a solar pond would have no advantages over a solid flat plate absorber. However, it was noticed that, in the case of some naturally-occurring ponds in Hungary, the water with the highest temperature was to be found at the bottom rather than at the top. The cause of this was traced to the high salt content of these ponds.

Many will recall, from experiments at school, that the amount of salts which can be dissolved in water increases rapidly with rising temperatures. This is not hard to understand. When particles from a solid crystal go into solution they do so because the forces between them and the molecules of the solvent liquid are greater than those forces which hold the particles together in the crystal. These forces, (sometimes called after F. LONDON (1930)), are not easy to determine from first principles, but solubilities are readily found by experiment. When the temperature is raised, the thermal oscillations of particles in the crystal help to overcome the forces which hold them together, so that it becomes easier for a particle to break away and to go into solution.

In the case of water the increase in concentration of certain salts held in solution helps to overcome the reduction in density due to expansion. A stable situation can be obtained in which the hotter fluid is also the denser and it remains at the bottom

84

of the pond. Thus, as well as a gradient of temperature with depth in the pond, there is also a gradient of salt concentration, which is lower near the surface than at the bottom. There will then be a diffusion of salt from the more highly concentrated zones towards those of lower concentration near the surface. It is necessary, for the maintenance of the highest temperatures, to remove salt from the surface and to re-introduce it at the bottom. The rate at which this must be done is quite low and does not present great technical problems. In experimental ponds in Israel, equilibrium temperatures have been found which approach 100°C.

We cannot go into further detail concerning solar ponds here, for the analysis of the heat and mass transfer in them is rather complex. The absorption of solar radiation is partly distributed throughout the body of the water and partly occurs at the base. There follows a complicated re-distribution of energy by conduction and by radiation exchange between different levels of fluid. As a result, the emission characteristics of the pond as a whole are different from its absorption characteristics. It is permissible, for present purposes, to think of its behaviour as not unlike that of a flat-plate collector in which the absorber plate has characteristics lying between the 'neutral' and the 'selective' properties used in our examples.

Solar ponds have a number of advantages over other collectors. They are by far the cheapest devices for collection of energy on a large scale, and they possess a large inherent energy storage capacity because of the high specific heat of water. Heat losses to the surrounding and underlying soil are more important in small ponds than in large ones, so that the most effective ponds are more than 50m square. In ponds of this size, disturbance of the surface by winds can affect the stability of the temperature and salt concentration distributions, so that wind-breaks have to be provided. However, these and other technical difficulties do not seem to preclude the use of solar ponds of virtually any size.

4.7 Useful consequences of absorption of solar energy

We have seen in this chapter how radiation from the sun is absorbed by the molecules of material bodies. One consequence of this, the increase in temperature of an isolated body, has been explored fairly thoroughly and ways of increasing the temperature rise have been examined. In succeeding chapters we go on to consider how this behaviour can be exploited to our advantage.

In Chapter 5 we see how we can use directly the temperature rise of bodies exposed to the sun, for example, in heating buildings and drying foodstuffs. These are fairly straightforward applications, which are already well in hand. Chapters 6 and 7 are concerned, respectively, with the transformation of energy from the sun into mechanical and electrical work. In both instances, the first stage of the process is the heating up of a body to act as a supplementary source from which energy can be drawn. It is found that the intermission of this heating stage causes limitations on the efficiency of the energy transformation process, these limitations being of a quite fundamental and unavoidable kind. They are, moreover, serious limitations. We shall find that, given tropical sunshine with intensity 800 W/m^2, we cannot expect to recover more than about 200 W/m^2 in the form of work. Therefore, in chapters 8 and 9, we consider energy conversion processes, respectively of a physical and a biological nature, in which there is no initial heating stage. We see that even here, high efficiency of energy conversion is not to be achieved easily, though the obstacles are of a different kind.

To obtain a working understanding of the processes involved in these systems, it will be necessary to examine further some of the thermodynamic and quantum-mechanical phenomena involved. In earlier chapters, we have already laid the foundations for this, but further development is necessary if an appreciation is to be gained of the limitations imposed by the nature of things on the behaviour of energy converting systems. The particularly serious kind of limitation mentioned above, for example, is embodied in the second law of thermodynamics, so we can expect to have to consider this law and its implications. As elsewhere, we shall take the simplest possible view of the physical basis of these matters. Before embarking on this, however, we can look first at some more straightforward uses of solar energy, in which the mere raising of the temperature of the system is the main requirement.

5

HEATING BY SOLAR ENERGY

He had been eight years upon a project for extracting sunbeams out
of cucumbers, which were to be put in phials hermetically sealed,
and let out to warm the air in raw inclement summers.

JONATHAN SWIFT (1667–1745)

ONE of the immediate results of the absorption of radiation
by a body is an increase in its temperature. This can be put
to use directly in meeting human needs, such as the heating of
homes and hospitals and the provision of hot water for domestic,
social and industrial purposes. Other straightforward applications of solar heating, such as the drying of foodstuffs and the
distillation of water, stem from great antiquity and form a broad
base upon which the use of solar energy for other purposes can
be built.

Perhaps because of the simplicity of these operations, devices
using solar energy directly have reached a higher level of development than those whose object is the production of power. The
solar water heater, for example, is already an accepted part of the
ordinary way of life in countries whose climate makes it economically attractive, such as Israel and Japan. In this chapter we
examine simple methods of predicting the performance of these
and other devices which make a direct use of energy originating
in the sun. It will be seen that modest extensions of existing
experience in this field could yield results of considerable
economic and social importance.

5.1 *The time required to reach equilibrium*
In the last chapter we examined conditions for obtaining a high
temperature at the points of equilibrium between the incoming
radiation and the losses from a body to its surroundings. The
time taken to reach this condition will depend upon a quantity
which we might call the THERMAL CAPACITY of the system.
This is the amount of energy which has to be absorbed to raise

the temperature by one unit. We saw in Chapter 3 that we can distinguish, for any substance, the SPECIFIC HEAT c, which is the energy required to raise unit mass of this substance by one unit of temperature, so that the total thermal capacity of a body of mass m is simply the product mc. Then it will be seen that, for any system, the total thermal capacity is simply the sum of the products mc of each of its parts.

As we have seen, a solid body stores energy partly in its electrons, but principally in atomic and molecular vibrations about their mean positions. The specific heat of the substance depends upon the number of atoms per unit mass and the number of ways these can vibrate. At ordinary temperatures, it is given approximately for many materials by the law of Dulong and Petit, which states that the product (specific heat × molecular weight) is a constant. In the units we are using here the constant has a value of about 7 Wh per kg per K. This value can be predicted quite well by using simple kinetic theory, to which reference was made in Chapter 3. We saw there that single atoms, when freely moving as in an ideal gas, would be expected to have a specific heat at constant volume of $3 R_1/2M_1$. This is associated with their kinetic energy. If, when vibrating about a fixed mean position, in a solid, we suppose that the atoms or molecules possess in addition an average potential energy equal to the average kinetic energy, we would expect the specific heat to be about $3R_1/M_1$. Thus the constant in Dulong and Petit's expression would have the value $3R_1$, which is almost exactly 7 Wh per kg per K. (A more precise description of the capacity of a substance to store energy, in particular its decrease at low temperatures, can be obtained by using quantum theory).

Now if we take as an example a flat plate solar collector in which the absorber plate is of copper ($c = 0.11$ Wh/kg K) 2 mm thick ($m = 14$ kg per m² of area) the thermal capacity of the plate alone is about 1.5 Wh/m² K. A typical glass cover plate would have a capacity of about 2.5 Wh/m² K. Then if the absorber and cover were raised by 100 K and 60 K respectively, the energy required would be 300 Wh/m². This corresponds to less than half an hour of noontide sunshine at tropical intensities. Even if we include the capacity of other parts of the system and the growing losses as the temperature rises, we may reasonably expect from this that the equilibrium temperature

88

would be reached in a typical case in a short period of time, perhaps less than an hour. This conclusion is reinforced by our common experience of the rapidity with which a stationary car heats up when standing in the sunshine.

Some rough computations will support this. To avoid using the calculus, which would allow us to deal directly with the changing conditions as the temperature rises, we will just divide the heating-up period into three intervals and suppose that during each part the rates of heat losses to the surroundings remain at the value corresponding to the average temperature during the interval. For this exercise, we will only look at the simplified case of a black plate with heat capacity $2 \cdot 5$ Wh/m^2 K exposed to solar radiation of intensity 800 W/m^2 as before, and losing heat by radiation at a rate $(T/64 \cdot 5)^4$ W/m^2. Convective losses of 4 W/m^2 K and a long-wave input of radiation of 200 W/m^2 will be assumed.

By the methods of section 4.4 we can show that if the plate is a neutral absorber, its equilibrium temperature would be about 70°C. If its initial temperature had been about 16°C, the average temperature for the first of the three equal temperature intervals of 18 deg C would be 24°C (297 K) at which the radiation and convective losses would be at a rate of about 480 W/m^2. Then at a nett input rate of 520 W/m^2, the plate would achieve the first third of its temperature rise in $(2 \cdot 5 \times 18)/520$ h, or about 5 minutes. However, the second third is found to take about 8 minutes and the last one about 23 minutes, giving a total time of about 40 minutes. (The large contribution of the last interval suggests that we would improve our accuracy by dividing this up further into smaller intervals. A more detailed analysis shows that in this example the plate is within about 1°C of equilibrium after an hour).

We might conclude, from this example, that the temperature of the absorbing plate would follow modest changes in solar intensity without much delay. It is evident that this will not be the case when the thermal capacity of the system is great, but it will simplify our later work if we can assume that most of the systems studied operate close to an equilibrium condition at all times. In most of what follows, therefore, we shall not complicate our rough analyses by making an allowance for the effects of the thermal capacity.

5.2 *Solar cookers and furnaces*

Perhaps the simplest direct use of solar energy is in the cooking of food, a universal process which is a great user of energy and manpower. A simple solar oven of the type shown in Figure 5.1 will quickly reach temperatures capable of cooking virtually any food over a period of some hours. The additional energy required to heat up the food and cook it (perhaps 300 Wh/kg) would often be as great as that required to heat the oven itself. But in a typical case, with the oven shielded from the wind, equilibrium temperature can usually be reached in little more than an hour.

More rapid cooking, and high-temperature processes such as frying, become possible if parabolic reflectors are used. Assemblies such as that shown in Figure 5.2, with mirrors about $1\frac{1}{2}$ m in diameter, have been employed in some numbers in various parts of the world. With rim angles of about 30° and quite crude production methods, effective concentration ratios of 500–1000 can be reached. The power delivered at the focus of such a device is between $\frac{1}{2}$ and 1 kW in tropical sunshine. With a cooking vessel about 15 cm across, shadowing of the mirror is negligible, but this kind of cooker needs to be correctly aligned with the sun and adjustments to its position must be made several times an hour. This is, however, a simple operation, within the competence of children or very elderly people for instance, since correct adjustment is obtained merely by keeping the shadow of the vessel near the centre of the mirror.

More elaborate methods for directing the mirror can be justified when the device is used for research purposes. A variety of large parabolic mirrors has been used for very high temperature work. With some of these, the sun's rays are directed onto the fixed mirror by a supplementary plane mirror, called a HELIOSTAT, which follows the apparent movement of the sun. The parabolic mirror then forms part of the fixed laboratory equipment, with a precisely located focus at which high temperature experiments can be carried out. The most famous centre for work of this kind, the Laboratoire de l'Energie Solaire, is operated by the French Centre National de la Recherche Scientifique in the Cerdagne region of the eastern Pyrenees near the border with Spain. Much pioneering work on the refining of highly refractory materials has been done there with the furnace

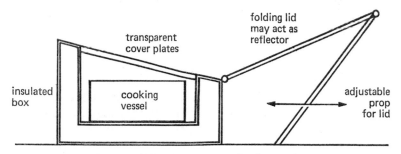

Fig. 5.1 'Hot box' type of solar cooker

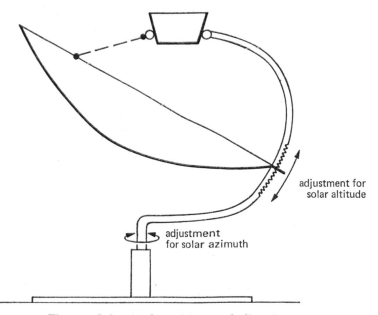

Fig. 5.2 Solar cooker with parabolic mirror

at Mont Louis, having a mirror of about 10 m in diameter. This is now being superseded at the new laboratory at Odeillo, nearby, where the parabolic mirror is no less than 50 m in diameter, made up of an array of some 8000 small mirrors, individually set to form the parabolic profile. This gigantic system delivers energy at rates up to about 1200 kW at a focal image of the sun nearly 50 cm in diameter. It is possible to undertake the refining of high melting point substances on a large scale with this equipment. Special rotating crucibles with water-cooled walls have been developed, capable of containing charges up to thousands of kilogrammes in weight. The incident energy melts a central zone of material which grows gradually outwards, kept in place by the rotation of the surrounding crucible. Only in a solar furnace can this be achieved without contamination from contact with the crucible walls— in this instance the central region is melted first. It has been possible to produce relatively large quantities of the most refractory substances, with melting points up to 3000°C. These include materials of great economic importance such as silica and zirconia (oxides of silicon and zirconium) which are required in the highest purity in several advanced technological fields which make use of their high melting points and chemical stability.

5.3 Heating of buildings and hot-water services

One of the most attractive direct methods for the use of solar energy is in the heating of buildings in cold climates or the provision of hot water for domestic use and in schools, factories, hospitals, and so on. Hot water systems based on the flat-plate solar collector are now being used on a large scale in Israel and Japan and quite large experimental installations for the heating of houses and swimming pools have been built in the southern U.S.A. and even in Europe.

The arrangement usually adopted is as shown in Figure 5.3. A fluid, usually water or air, in contact with the absorber plate of the collector, becomes heated and is induced to flow out of the collector by its buoyancy or by a pump in the system. The fluid then passes into a storage unit, from which it is withdrawn as required or in which is located a heat exchanger carrying the fluid which is to be heated. It acts in this respect like the familiar direct or indirect storage tanks of our domestic hot water systems.

Many refinements of this method are possible and here we are merely illustrating one of the simplest as an example, which we shall use to investigate the economics of such a system.

With the arrangement of Figure 5.3, the absorber plate acts

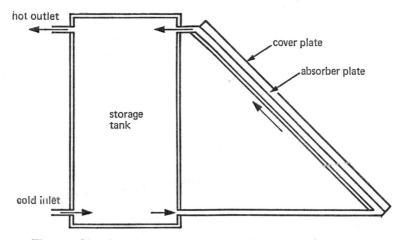

Fig. 5.3 Simple solar water heater with natural circulation

as a primary heat exchanger. The fluid either passes along the back of the plate or through a system of tubes attached to, or forming a part of, the plate. For air-heating collectors, a series of perforated absorber plates has been employed, with the air to be heated passing through the perforations. Where the rate of heat transfer between the fluid and the plate is good, as is usually the case with liquids, the absorber plate and the fluid will be at substantially the same temperature at any point. It is evident that the absorber plate will be cooler in the vicinity of the incoming fluid than near the outlet, the fluid having been heated in passing through the collector. The temperature rise of the fluid will depend upon its specific heat and the rate at which it passes. When the fluid is circulating under natural convection due to buoyancy forces, the flow rate may be low and temperature differences over the absorber plate amounting to several tens of degrees C can occur. When the fluid is pumped, on the other hand, the flow rate is greater, but the rate of heat transfer is increased. We shall, therefore, in most cases of interest have variations in temperature over the absorber plate, with attendant difficulties in analysis. However, since our purpose is to examine the principles from the simplest standpoint, we shall suppose that the absorber plate temperature can be taken to be substantially uniform and that the heated fluid leaves the collector at this temperature. The results will then apply mainly to systems in which the fluid enters at a temperature not much less than that at which it leaves. There must, of course, be some temperature rise as long as the fluid is acquiring energy.

With the above assumptions, we use a simple modification of the procedures of section 4.5 to determine the amount of heat which can be extracted from a collector at various fluid temperatures. As in previous examples, we will consider flat plate collectors with single cover plates and absorber plates with absorbtivities of 0·9 for solar radiation. We will recall that a collector designated 'neutral' will have an absorber plate with an emissivity of about 1·0 at its equilibrium temperature, whilst one designated 'selective' will have an absorber plate with an emissivity of about 0·1. These two cases may be taken to represent situations which are respectively easy to obtain and an upper limit to performance for simple systems of this kind. In both cases, we shall suppose that the cover plate is opaque to long-wave radiation and has the transmission characteristics described in Section 4.5. for solar radiation.

Now if the fluid in contact with the absorber plate maintains its temperature at T and in doing so extracts energy at a rate P_e, the energy balance for the absorber plate at equilibrium becomes

$$\alpha_s P = P_1 + P_e, \qquad\qquad 5.1$$

in place of equation 4.19. Here we must take P to be that part of the incident power which has passed through the cover plate. The equations for the interchange of radiation between absorber and cover and for equilibrium of the cover plate remain unchanged, being equations 4.18 and 4.20 respectively. We may then see by comparing equations 5.1 and 4.11, that for the simple conditions assumed, the extraction of energy at the rate P_e will result in an equilibrium temperature T which is the same as would occur with no extraction if the power incident on the absorber plate had been less than P. The temperatures would be the same, in fact, when the exchange P_1 has the same value.

That would be when the rate of energy reaching the plate (without extraction) has the value P_0 such that

$$a_s P_0 = P_1. \qquad 5.2$$

Then from equations 5.1 and 5.2, the common condition is when

$$a_s P_0 = a_s P - P_e$$

or $\qquad\qquad P_0 = - P P_e / a_s. \qquad 5.3$

If the equilibrium temperature of the absorber plate, T, is calculated for an appropriate range of values of input P_0 to the plate, we obtain the curves of Figure 5.4. by using equations 4.18 to 4.20. We may then use this figure to determine the power P_e which can be extracted at any given temperature for any value of the incident power P, by using the equivalence of equation 5.3. Some results of such an exercise are shown in Figure 5.5. It is found that the curves of this figure are almost straight lines and are well approximated for all values of input power P and temperature T by the relations

$$P_e = 0.9\, P - 5.8\, T \qquad 5.4$$

and $\qquad\qquad P_e = 0.9\, P - 3.5\, T \qquad 5.5$

for the neutral and the selective absorber respectively, when the unit of power is the W/m^2 and that of temperature the deg C in this case.

As well as showing the advantageous effect of the selective absorbing surface, Figure 5.5 illustrates the most significant feature of thermal power production from flat plate collectors. This is, that as the temperature of extraction increases, the fraction of the incident power which can be extracted must necessarily fall. This 'diminishing return' is inevitable because of the losses to the surroundings and raises a number of tricky questions concerning the best or optimum conditions under which these collectors should be operated.

Now that we can obtain the output from a given collector at any required temperature for any value of the intensity of solar energy, we can see how this will vary during the day, as the intensity varies. Although any collector could, in principle, be provided with a mechanism which would maintain it at all times facing the sun directly, this is rarely found to be justified

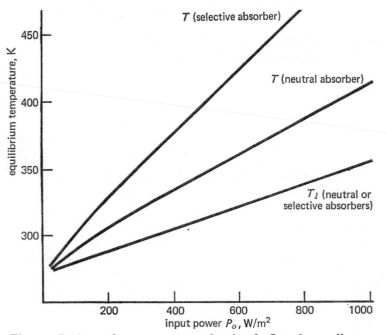

Fig. 5.4 Basic performance curves for simple flat-plate collectors

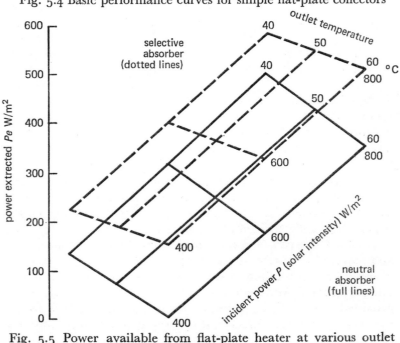

Fig. 5.5 Power available from flat-plate heater at various outlet
temperatures

economically and virtually all collectors used up to the present time have been stationary. We shall then proceed to obtain the output to be expected from a particular fixed collector system in a given location.

The first point to be settled is the best direction towards which the collector should face. We saw in Chapter 2 how to calculate the apparent path of the sun at any season in any latitude. In particular, we might recall here that the total variation in altitude arising from seasonal declination amounts to about 47° at solar noon throughout the year. Now the critical time for which the output of the collector is designed is likely to be the winter period when the demand on the system is highest but the output lowest.

At mid-winter, the solar declination is $-23\frac{1}{2}°$, but the coldest period of the year usually comes later, when the declination is somewhat less. Because of the thermal capacity of the land and ocean waters, which act as energy storage reservoirs, we find that the coldest period comes, in most parts of the world, some four to six weeks after the mid-winter solstice, (that is, in late January or early February in the northern hemisphere). Fixed collectors are therefore usually set up with a southward-facing aspect (in the northern hemisphere) and so inclined that the normal through the surface points somewhat above the highest noon solar point at mid-winter. In the nomenclature of Chapter 2, we shall then have, for the collector surface, $\varphi = 0$ and ψ greater than $(90-L-23\frac{1}{2})°$, perhaps by 10°. Then the inclination of the surface to the local horizontal will be about $(L + 23\frac{1}{2} - 10)°$. In the central U.K., for example, this would be about $65\frac{1}{2}°$, and for such a high latitude and above, it is often considered that a vertical collector surface would give nearly optimum results in winter time. The effect of variations in inclination of the collector on its output is not great when the rays are reaching the surface at angles close to normal, because the effective intensity is proportional to $\cos \theta$ (see section 3.5) which does not vary much with θ for θ close to zero. For the same reason, horizontal collector surfaces are satisfactory for regions close to the tropics, although the optimum angle might be some 30°.

We obtained values of total insolation for various latitudes in Chapter 2. In Figure 3.10 are given the solar intensities throughout the day for a location such as that of the central U.K.

($L = 52°$) in clear conditions. As a first example, we will calculate how much we could expect to recover from a simple flat-plate solar collector at this rather unpromising latitude. For this exercise, the effective solar intensity P at any hour and season is calculated by the methods of section 3.5, with allowance for the reduction due to the inclination of the surface and the transmissibility of the cover plate. These are shown in Figure 5.6. The total energy recoverable at any temperature can then be determined using the curves of Figure 5.5; some corresponding values are also plotted in Figure 5.6. The quantities are not inconsiderable. For example, we find that even at midwinter we could expect to recover from a selective collector 5 m square some 30 kWh per day at 40°C. However, these are in clear conditions, which so rarely are obtained in Britain. We shall return to the probable economic value of these devices later, and here we might merely note that in only a few instances so far has it been considered to be worthwhile actually to build solar heaters for buildings and swimming pools in Britain.

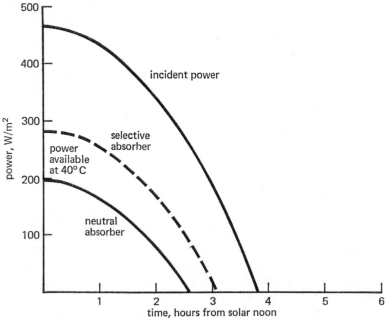

Fig. 5.6 Power available from flat-plate heater at various times. ($L = 52°$, winter solstice)

It was suggested in Chapter 2 however, that the most profitable place to exploit solar energy would be in the great desert regions around the tropics. It will be instructive to examine the probable output of flat-plate collectors in those regions. At the particular latitude of $23\frac{1}{2}°$, it is found by exercises such as the one above, that a collector with an inclination of about 37° to the horizontal ($\psi = 53°$) has virtually the same output at all times of the year. With the same assumptions as before, we obtain, for this situation, the results of Figure 5.7. We may see from this figure that each square metre of collector surface will yield about a megawatt-hour (1000 kWh) of energy in a year at temperatures which are quite suitable for hot-water systems for hospitals, schools and other institutions. The actual needs of a large hospital in the tropics for this kind of service might amount to about 500 MWh per year, which could, therefore, be provided very cheaply by flat-plate collectors with a total area of about 500 m². Space for collectors of this size would normally

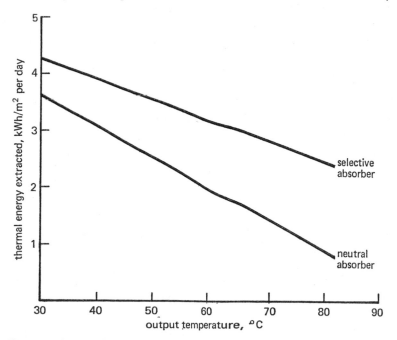

Fig. 5.7 Approximate output of solar water heater at latitude $23\frac{1}{2}°$ (any time of year).

be available on the roof of the buildings using the water, a generally convenient site which requires no additional ground space and in which the collectors perform the additional service of shading the building beneath.

5.4 Storage of heat energy

Storage is required in any hot-water system, to accommodate the varying demands upon it throughout the day and night. In solar-powered systems, larger storage facilities are required to allow also for the discontinuous availability of the sun's energy. Even in the very clear conditions we have been assuming in the previous examples, energy can only be extracted in useful amounts with suitable fluid temperatures for a very few hours on each side of solar noon. The higher the temperatures required, the briefer is the period; solar-powered house-heating systems, for example, demanding an output at about 60°C, have to rely on a receiving period of perhaps three hours only per day. So little of the demand is likely to coincide with the supply, that it is evidently necessary to be able to store enough energy for at least 24 hours in a manner which allows it to be recovered as heat at a useful temperature.

The requirements for hot water and for heating a domestic dwelling in a developed area with a climate such as that of Britain might amount respectively to about 15 and 150 kWh of energy per day in mid-winter. For a large hospital in the tropics, the hot-water requirement might be as much as a few MWh per day, all the year round. Now the specific heat of water is about $1 \cdot 2$ Wh/kg K, so that if the energy is to be stored in a quantity of water whose temperature is thereby raised by say 10 K, even one day's storage at the lower rate given above, and without losses, would need about 14000 kg of water, occupying 14 m³ of space. This is perhaps not too unreasonable, but the corresponding value for the hospital, say 200000 kg, would bring new engineering problems of its own.

A similar difficulty has had to be faced in the development of domestic night storage heaters, now quite popular in Britain. In these, an electrical element heats up a block of material during one or more off-peak periods when a cheaper rate can be charged for the energy. The heater must then store and gradually release this energy as required, between charging periods, without its temperature falling too far to be comfortable. In these, the conditions are arduous, in that the storage blocks have to be able to withstand the high local temperatures in the vicinity of the heating elements, so that refractory bricks are commonly used. Such heaters are thus rather massive and bulky.

With solar collectors, energy has been stored in underground water tanks and in chambers filled with boulders or pebbles. The latter are more suitable for hot-air systems, in which the air can be heated directly by passing through the spaces between the boulders. These voids occupy about a third of the total volume for roughly spherical boulders of uniform size. The large total surface area ensures good heat exchange characteristics in this case. Moreover, the pebbles only touch the walls of the chamber at a few points, reducing the heat losses to the surroundings. Their principal disadvantage is the low thermal capacity per unit volume—about a quarter of that of water.

In storage units employing the sensible heat of solids or liquids in this way, the energy is stored as increased kinetic and potential energy of the molecules of the storage medium. Much more energy is involved in changes of phase, where there is a change from an ordered structure to a disordered one. Melting and evaporation are examples of this. The increase in mean separation between the molecules in each case represents a substantial increase in potential energy, requiring a large energy input to bring it about. One substance which has been used in a solar heater is a paraffin wax which melts at about 55°C, with a latent heat of fusion of about 40 Wh/kg. This energy is given up again at this rather convenient temperature upon cooling. A storage of 150 kWh by this means requires only some 4 m³ of space. Certain hydrated salts have also been employed in this way. Glauber's salt, for example, $Na_2SO_4 . 10H_2O$, melts at about 32°C and dissolves in its water of crystallisation, taking in about 67 Wh/kg. Cooling to this temperature brings about a recrystallisation with liberation of the energy. In principle, this process can be repeated indefinitely, though it has been found that unless the melted salt is stirred, variations in concentration develop and recrystallisation of the whole is prevented. A continual search is in progress for other substances with indefinitely reversible phase changes at temperatures in the range 40 to 60°C, involving large latent heats. Many otherwise promising substances have had to be rejected through being expensive, explosive, poisonous, corrosive and so on.

5.5 Radiative cooling

Arrangements have to be made to cut off the flow to a flat plate collector when its temperature falls in the evening below that of the storage system. When the working fluid is water, it may also be necessary to cover the collector to avoid freezing. This can happen through the loss of energy by long-wave radiation to a clear sky after dark. It is possible to put this effect to good use in certain climatic conditions, providing a cheap cooling system.

Radiative cooling can only occur where the down-coming long-wave radiation from the atmosphere is substantially less than the radiation emitted by a surface at ambient temperature. As explained in section 4.4, this happens when the atmosphere is dry. In the presence of cloud, or when the air is humid, the atmosphere absorbs—and therefore emits—radiation almost like a black-body. Since the ground also behaves like a black-body, only slight differences in their respective temperatures can develop, arising mainly from the progressive transfer of radiation through successive layers of the atmosphere and ultimately out into space. In very dry climates, however, absorption by water vapour is weak or absent, the absorption spectrum of the atmosphere being discontinuous, with regions, known as *windows*, in which the atmosphere is virtually transparent. Some of the radiation from near the earth's surface can now pass unimpeded into space and the compensating downward radiation is much reduced.

Most of the long-wave radiation reaching the ground originates within the first few hundred metres of the atmosphere. It is to be expected that the intensity of this radiation will be related to the ambient temperature and the humidity near the ground. Although this radiation is not distributed with respect to wavelength exactly as for the black body, measurements show that it can usefully be expressed as a fraction of the black body radiation at ambient temperautre. An expression commonly employed is of the form

$$P_a = \sigma\, T_a{}^4(0.48 + 2.5\sqrt{w_a}). \qquad 5.6$$

The quantity w is a measure of the humidity, which we need not define now, as we shall deal with it more fully in the next section.

For the moment, it is sufficient to note that for very dry conditions, we may take the down-coming radiation to be about half the emission by a black body at ambient temperature.

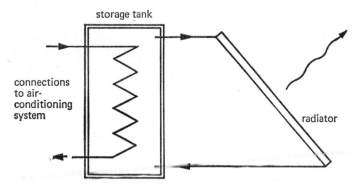

storage tank

connections
to air-
conditioning
system

radiator

Fig. 5.8 Simple radiative cooler with natural circulation

A radiative cooler takes the form of a solar heater, operating in reverse, as shown in Figure 5.8. Indeed, the collector of a heater system can be used during the night as a radiator, when connected to an alternative storage system. Since only long-wave radiation is involved, it is readily shown to be immaterial whether a cover plate is fitted or not. If we make similar assumptions to those made for analysis of the heater, the heat balance equation for the radiator in a steady state becomes

$$P_c + P_a = \sigma T^4 + h(T - T_a), \qquad 5.7$$

where P_c is the rate at which energy is transferred to the radiator by the fluid flowing through it from the storage system. As an example of the possibilities of radiative cooling, we might consider a case in which the ambient air temperature is 300 K (27°C). The cooler will be operating to reduce the temperature of a store of water, pebbles or a substance undergoing a change of phase, as discussed in the previous section. This, in turn, will be used the following day to provide a cool sink for an air-conditioning system. The energy extracted from the store at various temperatures under these conditions is shown in Figure 5.9. The variation is virtually linear with store temperature for this simple case. It is seen that when no energy is supplied to the radiator, its equilibrium temperature is just above the freezing point of water. We can assume that a practical storage system would be sufficiently large to operate at a fairly constant temperature and we may choose this temperature for convenience and economy. If, in our example, it is decided to keep the store around the average night-time air temperature of 300 K. we shall be able to reject over 200 W per m² of radiator area during the night. Then we shall be able to extract the corresponding amount from the cooled building during the day.

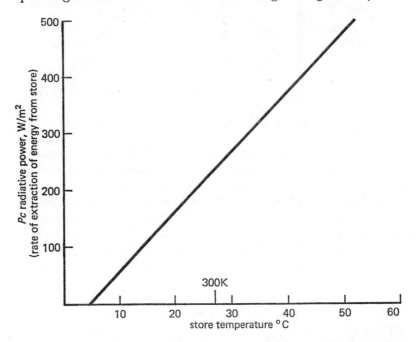

Fig. 5.9 Approximate performance of radiative cooler (very dry conditions)

It is evident that a radiative cooling system can be a very useful aid to comfort and efficiency where the meteorological conditions are right. Large buildings such as factories, schools and hospitals might be fitted economically with roof-mounted combined solar heating and radiative cooling systems, in which the same panels operate as collectors during the day and radiators at night, using separate stores. Such devices as these are particularly well suited to hospitals, where there is simultaneously a need for much hot water and for cool conditions in wards and operating theatres. The hypothetical hospital discussed in an earlier section, equipped with a 500 m² collector used for the equivalent of 8 hours per day as a radiator, could reject about ¾ MWh of energy during that period. This would be sufficient, for a hospital of this size, to maintain the air temperature in wards, theatres, and other essential areas at least 10 K. below the ambient temperature during the day time.

We shall return again to the question of cooling in the next chapter when considering refrigeration plant. In the present chapter we are dealing mainly with systems relying on radiative exchange processes and will look briefly at some more of these before examining the possibilities of power cycles.

5.6 *Distillation of water*
Water is scarcest in those places where sunshine is most abundant. It is not surprising to find, therefore, that solar energy has been used for many years in obtaining drinkable water by distillation from contaminated or brackish supplies. Many devices of varying complexity have been used for this purpose; one of the simplest systems is shown in Figure 5.10.

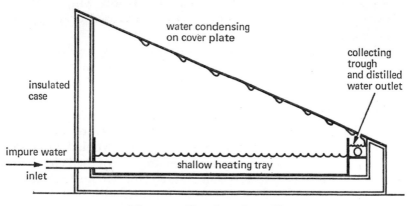

Fig. 5.10 Simple solar still

The unpalatable water is admitted into the tray at the bottom where it is heated by the absorption of solar energy. The base of the tray is usually blackened to facilitate this, since water is substantially transparent to the short-wavelength radiation from the sun. On occasions, the water itself has been rendered absorbent by the introduction of black dyes. As the temperature increases, the motion of the molecules becomes more vigorous and they are able to leave the surface in increasing numbers. Convection in the air above the surface carries them away and we have evaporation taking place (rather as it does from the surface of the sea). The rising air current, laden with moisture, cools on contact with the transparent cover plate and is unable to retain so much water, some of which then condenses on the plate and runs down into the collector channels at the edges. The cooled air returns to complete the convection current.

For the highest efficiency, it is desirable that the water should condense on the cover plate as a film. Dropwise condensation causes a significant fraction of the incoming sunlight to be reflected away again; even if the plate has a large slope, causing drops to run off easily, it is found that roughly half the surface is occupied by drops at any instant. Whether *dropwise* or *filmwise* condensation occurs depends upon the relative surface tensions of the water and the material of the plate. On glass thoroughly cleaned and free from greases, a film usually forms, but plastic materials, though much cheaper, are almost all unsuitable since they promote dropwise condensation. The reader will, perhaps, have noticed this property in the polythene film used for wrappers and bags.

Some recently-developed plastic films, which are capable of sustaining filmwise condensation, are at present nearly as expensive as glass, though a substantial demand would no doubt reduce the cost.

It is said that in the tropics, a solar distillation plant can generally be relied upon to produce distilled water at a rate equivalent to a rainfall of about $\frac{1}{2}$ cm per day. We shall attempt to justify this statement by a simplified analysis of the behaviour of a still like the one illustrated in Figure 5.10. A detailed analysis is a rather complicated problem in mass-transfer, but we can readily gain an adequae understanding of the process without going into too much detail.

We suppose that the water is in contact with an absorber plate and is at the same temperature T, as shown in Figure 5.11.

The cover plate is at the temperature T_1 as before. In our study of the flat-plate absorber, we expressed that part of the rate of heat transfer per unit area between two such plates which was due to convection in the form

$$q = h(T - T_1), \qquad 5.8$$

where h is the heat transfer coefficient, or rate of heat transfer per unit area of surface per unit time, per degree of temperature difference. Now we might represent this convection process, as shown in Figure 5.11, by two streams of air, each having a mass flow rate equivalent to m per unit area of surface per unit time, one moving outwards and one inwards.

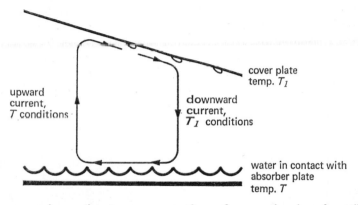

upward
current,
T conditions

downward
current,
T_1 conditions

cover plate
temp. T_1

water in contact with
absorber plate
temp. T

Fig. 5.11 Approximate representation of convection in solar still

The internal energy possessed by the air at any temperature T is cT per unit mass, if the properties are similar to those of an ideal gas, c being the specific heat of air. Then the warmer stream, leaving the lower surface, carries internal energy at a rate mcT, and the cooler stream carries internal energy at a rate mcT_1. Thus the nett rate of heat exchange between the surfaces by these streams is

$$q = mc(T - T_1). \qquad 5.9$$

Comparison of equations 5.8 and 5.9 shows that the rate of mass flow of air in these idealised convection currents would be given by

$$mc = h, \text{ or } m = h/c. \qquad 5.10$$

For example, the specific heat c of air is about 0·28 Wh/kg K, so that in a case for which h is 4 W/m² K, the rate of flow of air in the imagined air streams, m, would be 14·3 kg/m² h.

We now assume that the air convection currents move in the same way and at the same rate when they are laden with moisture. This kind of assumption is commonly made in elementary mass-transfer analysis but can only be justified when the mass-transfer rate is small. We shall further suppose that the air leaving each surface carries the amount of moisture appropriate to equilibrium at the temperature of the corresponding surface. When a gas is in contact with a liquid at a given temperature, some of the molecules of liquid leaving the surface later return again so that in a steady state as many molecules are returning per unit time as are leaving. Then the concentration of liquid molecules or vapour in the gas near the surface has a certain equilibrium value, known as the SPECIFIC HUMIDITY, w. The specific humidity, which is the mass of vapour per unit mass of gas, depends strongly on temperature, as may be seen from the graph of this property for water in air, Fig. 5.12. We are

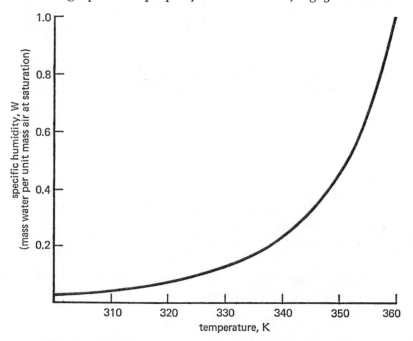

Fig. 5.12 Specific humidity for air at atmospheric pressure

supposing for present purposes that equilibrium is established rapidly enough not to be affected by the continual removal of some of the water molecules from the vicinity of the surface by the air convection currents, or their subsequent condensation on the cover plate.

Then, if we represent the convective process by the simultaneous movement of two air streams, each with a mass flow rate m per unit area of plates, the rate of transport of water outwards will be mw and the rate of transport of water inwards mw_1. The nett outward rate of transport of water is then $m(w-w_1)$; this will be the rate of production of distilled water per unit area of plates, M.

A liquid molecule has to be given a certain kinetic energy before it can escape from the attraction of its neighbours and join the vapour in the adjacent gas. At a given temperature, the molecules do not all have the same velocity (or kinetic energy) at any instant. The mean energy rises with temperature but there are always some molecules with more, and some with less energy than this. As the temperature rises, however, more of the higher-energy molecules possess enough to enable them to escape. (This process is very similar to that of thermionic emission from solids, which we shall examine in more detail in Chapter 7).

On the average, the energy required to escape from a body of liquid water is about 660 Wh per kg of water evaporated, so that the vapour carries a very substantial amount of energy. When the vapour reaches the cooler cover plate and part of it condenses there, this energy is communicated to the molecules of the cover plate. Thus, there now exists a powerful additional mechanism of heat transfer from the absorber plate to the cover plate. Its effect is so great, in fact, as to decrease the temperature difference between these plates in practice to about 10 K or less, in conditions for which the temperature difference in a flat plate collector, containing dry air only, would be about 40 K. We shall then have to amend the heat-transfer equations, 4.18 to 4.20, to allow for it. Only equation 4.18, giving the rate of energy exchange between the plates, needs to be changed, becoming

$$P_1 = h_1 (T-T_1) + \epsilon\sigma (T^4-T_1^4) + ml (w-w_1), \qquad 5.11$$

where l is the energy carried per unit mass of water vapour (the

LATENT HEAT of vaporization and condensation). This amend-
ment to the equation involves the additional assumption that the
presence of water vapour does not impede the radiative part of
the energy exchange.

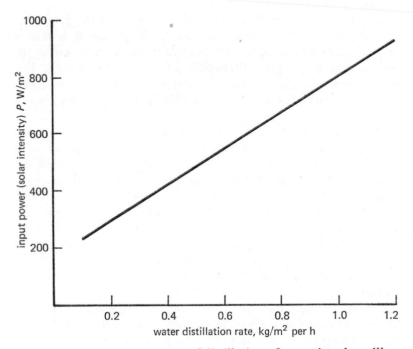

Fig. 5.13 Approximate rate of distillation of water in solar still

The solution of equations 4.19 and 4.20 with equation 5.11
must be done by trial and error, using the dependence of w on
temperature given in Figure 5.12. For an illustration of the
probable behaviour of a solar still, we may assume all the same
values to apply as we used for the flat-plate collector, taking
$\epsilon = 1\cdot0$ for the absorber plate and heated water. The results are
illustrated in Figure 5.13, showing them to give almost a linear
relationship between solar radiation input and water output,
well approximated in this example by the equation

$$M = (P - 160)/660, \qquad 5.12$$

when M is in kg/m²h with P in W/m².

To obtain the daily production from a given still, we need to know how P varies throughout the day. Then, using the data of Figure 5.13 or equation 5.12, the production at any time of day can be obtained. We then find, for instance, that for clear days in central Britain for which the value of P was shown in Figure 2.10, we could expect to distil as much as 7·6 kg of water per m² of still area a day in mid-summer, but virtually none at all in mid-winter. Corresponding values for a situation in a tropical zone with $L = 23\frac{1}{2}°$ are 8.2 kg/m² and 3.3 kg/m² per day. The average, over the year, is about 5 kg/m² per day. 5 kg of water occupy 5 000 cm³, so that over an area of 1 m² (10 000 cm²) the depth is the equivalent of $\frac{1}{2}$ cm of water per day, as quoted earlier.

The rate of production of palatable water from a still of the kind considered varies during the day, with the variation of intensity P. When the tray from which it evaporates is very shallow, the rate of production at any instant depends only upon the value of P at that instant. This is the case we have been examining. However, if the quantity of water in the tray is large, the temperature reaches an almost steady value after a few days, and water production then goes on steadily throughout the 24 hours. For this to occur, the amount of water present must be many times the daily production, say 100 kg/m², with a depth of about 10 cm.

One of the disadvantages of the tray-type still is the seasonal variation in its output. Much work has been done to overcome this difficulty. Among the other kinds of still investigated is one in which the water is evaporated from a sheet of dark-coloured, absorbent material which soaks up the water like a wick. This can then be inclined to the horizontal so as to receive the greatest radiation intensity and obtain the maximum output of water throughout the year. Another well-known type is the floating plastic still which forms part of the survival equipment of aviators and seamen of many nations. Much of the development work in this field over the past few decades has been associated with the name of DR. MARIA TELKES.

5.7 *Other applications of solar heating*
The direct heating action of the sun has been used for centuries in a variety of activities, many of which have important economic and survival values for the local community. Examples are

the production of salt by evaporation of saline waters and the drying of foodstuffs such as fruits and fish. Wherever a crop is highly seasonal, it is usually harvested in much larger quantities than can be utilised immediately, so that it must be preserved for later use. The removal of water upon drying produces conditions unfavourable for bacterial action and enables some foods to be stored without decomposition for inter-seasonal periods of the order of a year.

Solar drying is usually a slow process and its rate may be the dominating factor in determining the output of a given local industry such as salt production, seasoning of wood, curing of rubber and so on. Any increase in rate of drying would directly induce an increase in productivity for the enterprise as a whole. This is an area in which important developments are likely. Some studies have already been made into possible methods of enhancing the ordinary drying effect of exposure to the sun and into the replacement of conventional methods of heating by the use of solar energy. An example is the work at the National Physical Laboratory of India, showing how the use of simple solar concentrator mirrors could speed up the evaporation of water from palm and sugar cane juices, the source of oil and sugars for rural communities.

We can readily estimate the effects of using plane concentrators with a simple drying frame for foodstuffs, as illustrated in Figure 5.14. With some care, a concentration ratio of about 2 is possible with this arrangement, employing polished metal reflectors. To obtain this degree of concentration throughout the day, however, it is necessary to adjust the orientation of the frame, perhaps twice an hour, to allow for the sun's apparent motion.

If we again relate the rate of mass transfer for moisture to the rate of heat transfer due to buoyancy, we would write the heat balance per unit area for this system as

$$P_\mathrm{a} + a_\mathrm{s}P \times \mathrm{CR} = \epsilon\sigma T^4 + h(T - T_\mathrm{a}) + ml(w - w_\mathrm{a}), \quad 5.13$$

with $m = h/c$. Here we are assuming, for an approximate analysis, that the surrounding air and the air in contact with the surface of the drying material is always saturated at the respective temperatures. P_a is the unconcentrated radiation, consisting of the diffuse part of the solar radiation and the long-

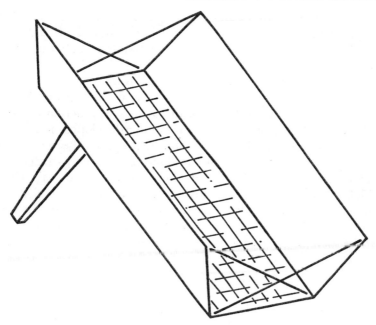

Fig. 5.14 Drying frame with flat concentrators

wave radiation from the atmosphere which might typically
amount to a total of 300 W/m², with an ambient temperature
T_a of 300K (27°C). With these values, we find that when the
direct solar radiation intensity is 800 W/m², the rate of evaporation of water, which is $m(w - w_a)$, is about 0·5 kg/m² per h
without a concentrator and 1·4 kg/m² per h with CR = 2. Over
a whole day's operation, the rate of drying is increased by a
factor of about $2\frac{1}{2}$ when the concentrator is used.

These results are for a drying frame used in still air. Since the
rate of drying is related to the heat transfer to air passing over
the dryer, it is augmented when a wind is blowing. The gain is
not, however, proportional to the increase in heat transfer
coefficient, because the dryer operates at a lower temperature
in a wind, and the specific humidity at saturation is then lower.
The effect of a low temperature on the rate of evaporation is
shown by comparing the results of this section with those of the
previous section for a solar still. It is evident that more effective
drying can be obtained by employing a cover plate, as with the

still. The advent of cheaper transparent films with suitable properties and durability for covering materials would undoubtedly make greater drying rates possible in many circumstances.

Substantial progress in the field of solar drying is to be expected from the work in being in several parts of the world. Moreover, this will probably lead to the introduction of new industries into remote and hitherto underdeveloped areas and to the adaptation, to the use of solar heating, of other processes involving drying, such as the production of paper, cardboard and constructional strawboard. Many processes of this kind which might be so adapted, however, could only be carried on where mechanical or electrical power is available also. In the next chapters we shall consider ways in which the sun's energy can be used in the production of these more conventional forms of power.

6

CONVERSION OF SOLAR ENERGY INTO WORK

As though a rose should shut, and be a bud again.

JOHN KEATS (1795–1821)

So far we have only been concerned with the conversion of the sun's energy into the form we identify as HEAT and using it in that form. If we want to use the energy in driving machines we shall need it in the form of WORK. One kind of machine that could be driven by a source of work is the electrical generator, which provides energy in a convenient form which can be transported over long distances and converted at its destination into heat or into work.

It would be natural to suppose at first that we could convert all of the sun's radiant energy into work, if it were not for inevitable frictional losses which reduce the efficiency of our machines. However, it is found that, where the entry of the energy into the machine is through a heat-transfer process, there is a fundamental limitation on how much of it can be recovered as work in a continuous process. This applies, for instance, where the energy is obtained by the absorption of the sun's radiation or by the burning of a fuel as in internal combustion engines; in both these cases, the energy on entry to the system is in the form of heat. It must be stated that *even if we had a perfect machine, without mechanical losses of any kind, we could still only recover a part of this energy as work.* This is so unexpected a result if it is met for the first time, that we shall need to spend a little space in the first part of this chapter in examining it. We find, however, that if it were not so, it would be as astonishing as the rejuvenation of the rose in the quotation which heads this chapter.

In the remainder of the chapter, we shall consider the probable performance of simple combinations of solar energy collector and heat engine, with a view to estimating the economic value of devices subject to such a fundamental limitation on their efficiency.

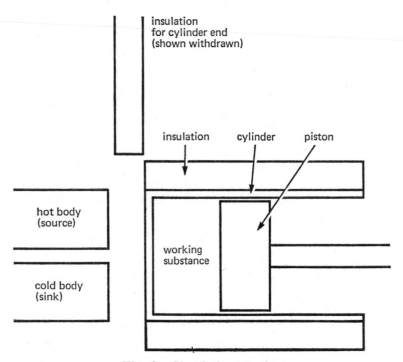

Fig. 6.1 Simple heat engine

6.1 *The first law of thermodynamics*

Suppose that the simple cylinder and piston shown in Figure 6.1 can be taken to represent any machine for producing work continuously. In this case, we imagine an output shaft being caused to rotate as the piston moves in and out. Such a device is called an ENGINE. We are going to assume that we can apply to all engines certain deductions about the behaviour of this one, which we have chosen to consider because it is one of the simplest we can imagine. So that the thermodynamic behaviour can be described as simply as possible, we consider the space within the cylinder to be filled with an ideal gas, whose properties we derived from first principles in Chapter 3. In general the fluid within an engine is called the WORKING SUBSTANCE.

We need not specify the nature of the surroundings except that we will suppose that the pressure there is negligible so that there is no restraint offered to the motion of the piston by forces applied to its exposed surface.

Now if we bring a hotter body into contact with the cylinder, energy will be transferred as heat and will ultimately increase the internal energy of the gas. To simplify matters, we will suppose that no energy is retained by the cylinder wall. If a quantity Q of energy enters and nothing else is allowed to happen, the internal energy of the gas will increase by an amount ΔU_1, where

$$Q_1 = \Delta U_1. \qquad\qquad 6.1$$

According to our ideas about the ideal gas, an increase in internal energy means an increase in temperature, and according to equation 3.7, an increase in pressure. The hot body is now removed, and replaced by an insulating blanket which prevents any further heat transfer across the cylinder wall in either direction. (A non-conducting envelope like this is called ADIATHERMAL). If the piston is then allowed to move outwards under the action of the increased pressure, it will cause work to be done on whatever it is pushing against, say a machine to be driven. This work can only be obtained at the expense of a reduction in the internal energy of the gas, since we are allowing no heat transfer across the cylinder wall. We could permit the piston to move out until the internal energy had fallen to the value it had before the initial heating began. In this process an amount W_1 of external work is done. Then all the increase in internal energy due to heating will have been converted into work, for we shall have $W_1 = \Delta U_1$ and hence $W_1 = Q_1$. (However, we must not think that this is a limit. It is easy to see that if we did not stop the piston when this condition had been reached, we could continue to obtain work, though in doing so we would cause the internal energy of the gas to fall below its original value).

The trouble with this situation, which seems at first to be contrary to the opening remarks of this chapter, is that the system is not now in its original state. The gas has been restored to its original internal energy, but the piston is not in its original position. If we want to obtain work *continuously*, we shall have to restore the system to its original state in every particular, so that the process can be repeated. That is, we must make the system go through a CYCLE. It is when we introduce cycles that we encounter the limitations. The original state cannot be restored simply by pushing the piston back in, for this would require work to be done in compressing the gas, which would then have more internal energy at the end of the process than it had originally. We conclude that, to restore the original state completely, we shall have to extract internal energy from the gas, so that it is at a lower temperature, before compressing it again. To do this, we could remove the insulation, bring up another body, cooler than the cylinder, and allow energy to be transferred to it as heat. We must take out an amount $Q_2 = \Delta U_2$, such that, in restoring the piston to its original position, an amount of work $W_2 = \Delta U_2$ has to be done. Thus, for the cycle as a whole, we have introduced a nett amount of energy $(Q_1 - Q_2)$ as heat and obtained a nett amount of energy $(W_1 - W_2)$ as work. Since we have arranged that there has been no overall change in the internal energy, we must have that

$$(Q_1 - Q_2) = (W_1 - W_2). \qquad 6.2$$

This equation embodies one form of what is called the FIRST LAW OF THERMODYNAMICS, which states that if any system is taken round a cycle so that its original state is completely restored, the nett amount of energy transferred to it as heat is equal to the nett amount of work extracted from it. This is to say, of course, that energy is *conserved* in the cycle. Though the idea had long been in use, this formal statement was first given in 1908 by POINCARÉ, who also framed the corollary which is often substituted for it, namely that in any process in which an amount of heat Q is transferred and an amount of work W is extracted, the balance represents the change of internal energy of the system, i.e.

$$Q - W = \Delta U. \qquad 6.3$$

This equation, as a definition of internal energy, is often used as starting point for the development of thermodynamics without reference to the microscopic kinetic behaviour through which we have approached it in Chapter 3 and the present chapter.

In these processes, it seems at first that heat and work are indistinguishable. We recall that the English physicist JOULE performed in the years following 1840 a series of experiments in which a given temperature rise—and hence a given change in internal energy for a simple system—was produced either by heating or by mechanical work done on the system by stirring. Most of us have repeated a form of his experiments in which the temperature is raised by frictional work or by passing an electric current through a resistive wire. Joule's experiments established an equivalence between the units of heat and work then in use, the constant of proportionality being called the 'mechanical equivalent of heat'. Nowadays, we do not make this distinction and use the same unit for both, justly named the JOULE. (In this book, of course, we are using the more convenient and familiar unit, the kWh, which is simply 3·6 million joule).

6.2 *The second law of thermodynamics*

When we return to consideration of our elementary cycle, we must not imagine that, in view of equation 6.2, we have a perfectly efficient system for converting heat into work. For we must recall that the energy Q_1 was introduced as heat by a process equivalent to bringing up an external body into contact with the cylinder. From the discussion of Chapter 3, we know that for heat transfer to take place, this body must be hotter than the gas at all times. Then, after the piston had moved out, we brought up another body to reduce the temperature of the gas before we recompressed it. This body must be cooler than the gas at all times. It follows that the first body (called the SOURCE) and the second (called the SINK), *cannot be the same body*. Here we have encountered, in the simplest possible example, one of the great principles of science, known as the SECOND LAW OF THERMODYNAMICS. This declares it to be a natural law of our particular universe that all cyclic processes for producing work, of whatever complexity, have a characteristic in common with our example: they must have both a source and a sink. If we have noticed that all power stations, large factories and so on have cooling towers or other devices in which heat is being rejected into the atmosphere or a river as a sink, we are already prepared to accept the law in this form. It seems to have occurred first to the French engineer SADI CARNOT in 1824 as a result of observations on the characteristics of steam engines. Formal statements and corollaries of the law with successively greater clarity and generality, were given by CLAUSIUS (1850), KELVIN (1851), PLANCK (1897) and POINCARÉ (1908). For our present purposes, we may regard as definitive the statements of Kelvin and Planck to the effect that *it is impossible to construct a device which will operate in a cycle and perform work, whilst exchanging energy in the form of heat with a single reservoir.* However, we see also that in our example, the necessity for both a source and a sink is due to the fact that *heat transfer can take place spontaneously only in one direction: from a hotter to a cooler body.* This is the basis of the familiar form of the second law given by Clausius.

We also can see in our example an essential difference between the roles of heat and of work as agencies of energy transfer. When we wanted to increase the internal energy of the gas initially, we could have done this either by heating or by doing work through a stirrer inserted into the gas. However, when we later wanted to *reduce* the internal energy, we could only do it by cooling—there is no way in which it could be extracted as work without changing the shape of the boundaries. Thus, there are certain limitations on what changes can be brought about by work interactions alone. Another form of the second law based on this is due to CARATHÉODORY (1909) and is sometimes preferred as a more general starting-point for the development of thermodynamics.

The implications of the second law are not as widely known as they should be, perhaps because no-one has succeeded in formulating a single statement of the law which covers all that we require of it. The amended Clausius form given above is preferred by some because it seems to represent, without the formal apparatus of engines and cycles, the essentially irreversible nature of the universe. Buildings crumble; men die; hot bodies cool; roses never return to the bud.

6.3 *The efficiency of cyclic processes*
The object of our study is to see how much of the solar energy could be recovered in the form of work if it is caused to operate a heat engine. We must not stray too far into detail, but the limitations on this process are so important that we must examine them a little further.

If we think again of our simple cycle we shall be able to see at once that the thermodynamic limitations, which are embodied in the second law, must make the cycle less than perfectly efficient. We had to supply the energy Q_1 as heat from the source; the energy Q_2 is heat rejected to the sink which cannot be used to supply energy to the next cycle because the sink temperature is low. If an amount Q_2 of the energy supplied Q_1 is, in effect, wasted in this way, the energy employed usefully is $(Q_1 - Q_2)$ and hence the EFFICIENCY, η, which is the fraction of the energy supplied which is usefully employed, is

$$\eta = (Q_1 - Q_2)/Q_1 = 1 - Q_2/Q_1. \qquad 6.4$$

Alternatively, we note that the nett useful work obtained during one cycle is $(W_1 - W_2)$ so that the efficiency is also given by

$$\eta = (W_1 - W_2)/Q_1, \qquad 6.5$$

which, by equation 6.2, is evidently the same result.

It must be emphasised again that the loss of efficiency is inherent in the nature of things and is not a mechanical deficiency associated with the mechanism used to realise the cycle in practice. Accepting that there must be a limitation, then, we next want to know what would be the highest efficiency it is possible to get with a 'perfect' engine. It is found that such an engine must possess a characteristic known as REVERSIBILITY. The features of this characteristic are shown in Figure 6.2. When the engine, shown in the left-hand half of the figure, is reversed, the energy flows all take place in the opposite direction. In the reversed condition, work must now be supplied to the engine and energy is obtained in the form of heat from the sink and delivered at the source. These are the characteristics of the REFRIGERATOR and the HEAT PUMP, to which we shall return in a later section. It is easy to show that an engine which could be reversed in this way would be the most efficient that could be made. For suppose that we imagine one more efficient than this, used to drive our engine in the reversed condition, as shown in Figure 6.3. To provide the work W, the super-efficient engine would need to extract only an amount Q'_1, less than Q_1, from the source. It follows that the energy discharged into the sink, $(Q'_1 - W)$, is also less than that drawn from it by the reversed engine, $(Q_1 - W)$. It is then seen that the nett energy drawn

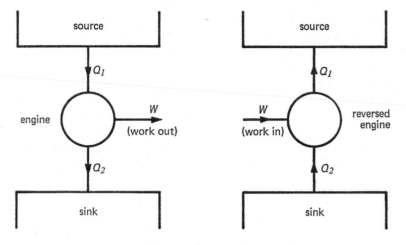

Fig. 6.2 Energy flows in heat engine

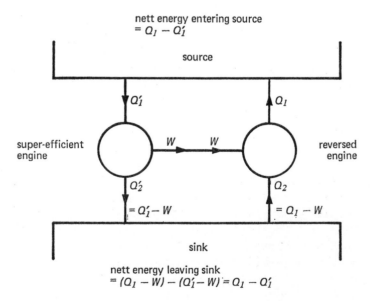

Fig. 6.3 Showing impossibility of engine more efficient than reversible one

from the sink is ($Q_1 - Q'_1$) and the nett energy delivered to the source is also ($Q_1 - Q'_1$); no nett work has to be supplied. If this could happen, it would contravene the Clausius statement of the second law. Without any outside intervention, energy ($Q_1 - Q'_1$) would have passed from the cooler to the hotter body. We must conclude then that we could not, in fact, obtain an engine more efficient than the reversible one.

Since we have said nothing about our ideal engine except that it is reversible, we must conclude also that *all* reversible engines working between the same source and sink would be equally efficient, regardless of their mode of operation, working fluid, etc. What then, does this ideal efficiency depend on? The only other characteristics of the system are the temperatures of the source and sink, so we must conclude that *the efficiency of an ideal reversible engine depends only on these temperatures*. Thus, following equation 6.4, we could write

$$\eta = 1 - Q_2/Q_1 = 1 - f(T_1, T_2), \qquad 6.6$$

where $f(T_1, T_2)$ means some function of T_1 and T_2 only, whose form we have not yet determined. Now consider the arrangement shown in Figure 6.4. Two reversible engines are operating

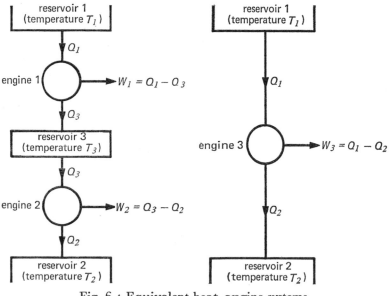

Fig. 6.4 Equivalent heat engine systems

between three reservoirs. If we arrange them so that the lower engine extracts just as much energy from the central reservoir as the upper engine discharges into it, this system is just equivalent to a single reversible engine producing a total amount of work $(Q_1 - Q_3) + (Q_3 - Q_2)$, which is $(Q_1 - Q_2)$. Then, by equation 6.6 we would have for each engine in turn

$$Q_3/Q_1 = f(T_1, T_3),$$
$$Q_2/Q_3 = f(T_3, T_2)$$
and $\qquad Q_2/Q_1 = f(T_1, T_2),$

where the form of the function f is to be the same in each case. Now since $Q_2/Q_1 = Q_2/Q_3 \times Q_3/Q_1$, the requirement is that f shall satisfy the relation

$$f(T_1, T_3) \times f(T_3, T_2) = f(T_1, T_2). \qquad 6.7$$

It is now known that more than one function will satisfy equation 6.7, but the simplest, (which is used because it can also be shown to be consistent with other thermodynamic considerations), is that

$$f(T_a, T_b) = T_b/T_a. \qquad 6.8$$

That is, equation 6.7 becomes

$$\frac{T_3}{T_1} \times \frac{T_2}{T_3} = \frac{T_2}{T_1}.$$

Thus we come to one of the most important relations in our study. We have shown that the greatest possible efficiency of any engine producing work as a result of heat exchange with a source and sink is, by equations 6.6 and 6.8,

$$\eta = 1 - \frac{\text{sink temperature}}{\text{source temperature}} \qquad 6.9$$

and that the energy flows therefore satisfy the equality

$$Q_2/Q_1 = T_2/T_1,$$
or $\qquad Q_1/T_1 = Q_2/T_2. \qquad 6.10$

The efficiency given by equation 6.9 is fittingly called the CARNOT EFFICIENCY.

Equation 6.9 shows up sharply the limitations placed by the

nature of things on our use of heat engines. We can increase the efficiency by increasing the source temperature or decreasing the sink temperature. But we are not at liberty to do either without limit. The first is limited in practice by the temperatures which can be tolerated by the materials of which the engine is made. On the other hand, there are no natural perpetual sinks with temperatures much below the atmospheric temperature. Those readily available to us are the air itself or cooling water from wells, rivers and the seas.

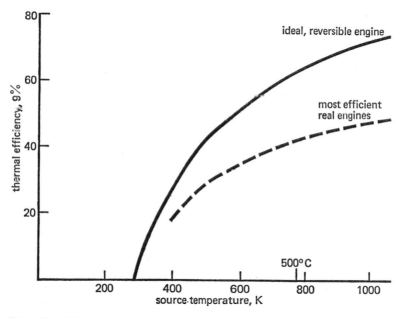

Fig. 6.5 Efficiency of heat engines with sink temperature 290K (17°C)

6.4 *Mechanical power generation by heat engines*

The foregoing results have profound implications for the question of power generation from solar energy. If we suppose that natural sinks in tropical regions will not have temperatures lower than about 290 K (17°C), the highest possible efficiency of any heat engine, as given by equation 6.9, is shown in Figure 6.5 for a range of source temperatures. These are for ideal engines having no imperfections; we see that even if metallurgical limitations would allow us to operate with source temperatures of 1 000 K, the maximum efficiency would be about 70%. However, with the utmost care in mechanical design, we have so far been unable in practice to devise an engine whose actual efficiency is greater than about two-thirds of the Carnot efficiency. The overall efficiency of real engines is thus around the values shown by the dotted curve in Figure 6.5. To understand this, we need to know a little more about the ideal or reversible cycles which the full curve represents.

A reversible cycle is one which can be made to proceed in a reverse direction, in such a manner that the working fluid and the surroundings are in precisely the same states at any given point in the cycle, whichever way the cycle proceeds. In the reversed cycle, the direction of the energy transfers, as heat and work, are evidently the reverse of those in the forward cycle. We shall return later to a consideration of the usefulness of reversed engine cycles.

Meanwhile, we can soon see that our simple cycle is not reversible. It is convenient to represent a thermodynamic cycle by diagrams in which the coordinates are the pressure of the gas and the volume occupied by it. (We recall from Chapter 3 that if the gas is ideal, these two quantities also determine the temperature). Our simple cycle is shown in this form in Figure 6.6. If point 1 represents the initial condition, energy is first added as heat, raising the pressure to p_2, then the piston is allowed to move out, doing work so that the pressure falls to p_3. Heat rejection to the sink then lowers the pressure to p_4, so that it can be restored to p_1 again by recompression. Now if we begin at point 1 and first allow the piston to move to the *right*, we might cause the path 1–4 to be followed reversibly, becoming an expansion instead of a compression. There are some difficulties about doing even this, but we will pass over them here, for there follows an impossibility. We cannot cause the section 4–3 (and later 2–1) to be traversed with all the states the same as they were for the original cycle. This is because energy cannot be taken in as heat from the sink (or delivered as heat to the source) at a point where the sink is cooler (or the source hotter) than the working fluid.

Heat transfer can only take place in one way. In our cycle, the source temperature is at least as great as T_2 and the sink

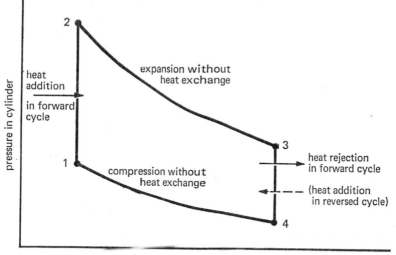

Fig. 6.6 Thermodynamic cycle for simple heat engine

temperature at least as low as T_4, so that using equation 6.9, the Carnot efficiency of a truly reversible engine using this source and sink would be $1 - T_4/T_2$, whereas the efficiency of one using our cycle is evidently less than this. (It can be shown in fact to be $1 - T_3/T_2$).

The irreversibility—and consequent loss of efficiency—of most practical engine cycles is due in large measure to the heat-exchange processes, as in our example. Carnot's original conception of a reversible cycle, illustrated in Figure 6.7, includes heat-exchange processes along the section 1–2 and 3–4 in which the working substance is at constant temperature (the temperature of the source and sink respectively). These processes do not take place at constant volume. The work extracted along 1–2 is made exactly equal to the energy added as heat so that the internal energy (and hence the temperature) remains constant. Since heat transfer can only take place if there is a temperature difference, it is evident that the temperature of the working substance cannot be exactly equal to that of the source along 1–2 or the sink along 3–4. It follows that the cycle can only be imagined to approach true reversibility as these temperature differences are reduced. If they were actually zero, the cycle could be traversed in either direction. In this imagined limiting condition, the cycle has the Carnot efficiency.

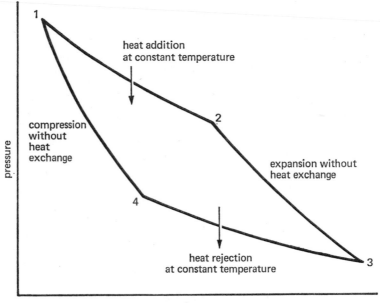

Figure 6.7 Thermodynamic cycle for Carnot reversible engine

Another important cause of irreversibility is *friction*, for example, friction between a piston and the cylinder wall. This always opposes motion, so that if an attempt is made to reverse a cycle, the force in the piston rod is not the same at a given point in the cycle when the piston is moving in one direction as it was when it moved in the other. Friction may also occur between the molecules of the fluid or between the fluid and solid boundaries.

In real engines, also, it is never possible to construct truly insulating or adiathermal boundaries, which are a feature of many idealised reversible systems. Real engines are thus inevitably irreversible and approach the Carnot efficiency only approximately, however ingeniously they are contrived. In the heat engine most of us know best, the internal combustion engine, the working substance follows a cycle something like our simple one (Figure 6.6), which is called the OTTO CYCLE, Figure 6.8. There is, however, no source as such, the energy increase along 1–2 being brought about by the burning of a fuel in the air which has been first compressed along 4–1. After expansion in the working stroke 2–3, the gas is discharged at 3, since it cannot be taken round the cycle again, its oxygen having been depleted so that it would not support further combustion. The discharge of this hot gas and the admission of fresh cold air at state 4 corresponds to heat rejection to a sink—in this case the atmosphere.

In using solar energy, or any external source, we could operate in a closed cycle, using the same body of fluid all the time. A cycle well adapted to this is known as the STIRLING CYCLE, and is shown in Figure 6.8. As in the Carnot cycle, heat transfer occurs along 2–3 and 4–1 at constant temperature, a corresponding quantity of work crossing the boundary at the same time. Along 1–2, the working substance is imagined heated by contact with a series of sources of gradually increasing temperature, and along 3–4 cooled by contact with a similar series of sinks. Since the energy exchanged and the temperature change along 1–2 is the same as along 3–4 it is possible to arrange for the sources and sinks to be the same bodies. The heat rejected in one part of the cycle is returned in the next, being stored meanwhile in these bodies, which constitute the REGEN- ERATOR. Over one complete cycle, the state of the regenerator is completely restored, so that it is not regarded as part of the surroundings, but as part of the engine. In practice, a regenerator is a matrix of fine tubes or wire gauze. If it is sufficiently large, so that its thermal capacity is much larger than the energy exchanged per cycle, the temperature of its various segments remains almost constant. Then the gas in contact with it exchanges heat with a body which is at all times at nearly the same temperature as itself. The process is almost reversible and the efficiency high. A well-known hot-air engine, based on the Stirling cycle, is marketed by the Philips company of Holland, and others have been built by the General Motor Corporation of America. Efficiencies as high as 70% of the appropriate Carnot efficiency have been reported for engines of this kind. The ingenuity with which the ideal Stirling cycle has been

Otto cycle

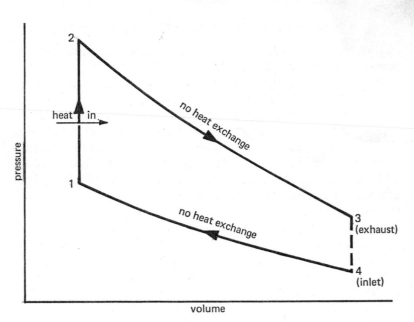

Open cycle resembling that of I.C. engine

Stirling cycle

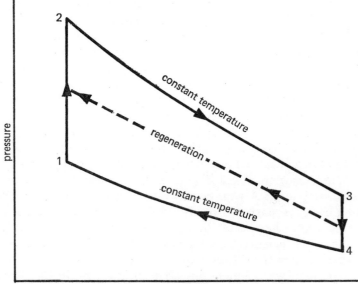

Closed cycle resembling that of hot-air engine

Fig. 6.8 Ideal thermodynamic cycles

realised in these engines is illustrated by the sketches of Figure 6.9. A displacer piston is used to cause the working gas to pass through the regenerator at the appropriate points of the cycle, a separate power piston being used for the compression and expansion processes. A number of engines of this kind has already been built and operated entirely by solar power, so that their construction and use for this purpose can be regarded as proven.

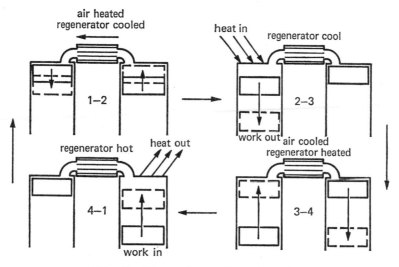

Fig. 6.9 Two-cylinder hot-air engine

Another regenerative cycle, but one in which the regeneration occurs at constant pressure rather than at constant volume, is the ERICSSON CYCLE, shown in Figure 6.10. We need not confine our thinking to reciprocating engines with pistons and cylinders. In Figure 6.10 is also shown a possiblearr angement for realising the Ericsson cycle, in which the expansion along 2–3 takes place in a TURBINE and the compression along 4–1 in a COMPRESSOR, mechanically coupled together. Now the fluid flows steadily and continuously round a closed system. Practical difficulties of realising this cycle and the Stirling cycle occur in the design of suitable regenerators and of devices in which expansion and compression can take place at almost constant temperature with the associated heat exchange.

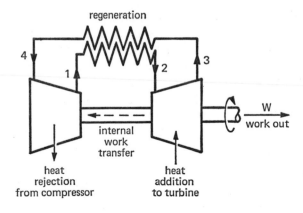

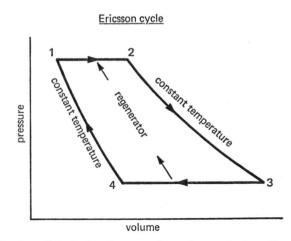

Ericsson cycle

Fig. 6.10 Ideal closed gas cycle and practical realisation

Engines using fluids which evaporate and condense at various points in the cycle are in universal use, and have been since the invention of the steam engine.

Figure 6.11 shows a schematic arrangement of such a system similar to the steam-turbine plant generally used for electrical power production on a large scale. In this case, a solar heater takes the place of a boiler which is conventionally heated by the products of combusion or by a nuclear reaction. Systems such as these have undergone extensive development and, at the expense of considerable complexity and refinement, modern plants are operating at overall efficiencies approaching 40%, representing over 60% of the appropriate Carnot efficiency.

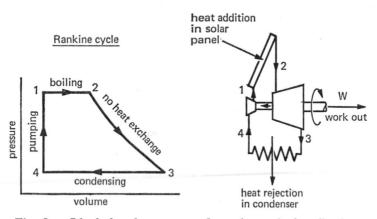

Fig. 6.11 Ideal closed vapour cycle and practical realisation

Many ingenious variations of thermodynamic cycles and means of realising them in practice have been devised during the last century. It would be inappropriate to consider their relative merits here, though enough has been said to indicate some of the possibilities and difficulties of heat engine development. We began by wishing to know how much of the energy supplied to a heat engine can be made available as work. In the previous section, we saw that we could not, by any means whatever, obtain more than the fraction representing the corresponding Carnot efficiency. From the present section, we might conclude that it would be unwise, with any simple system, to expect in practice to recover more than about half of this quantity. We shall assume this to be the case as we come to consider the possibilities of power generation by engines using solar energy.

6.5 *Combined engine-collector performance*

The combination of an engine with a solar energy collector presents further difficulties. We have just seen that a high maximum temperature in the cycle leads to a high engine efficiency. But where a flat-plate collector is used, we saw in Section 5.3 that the efficiency with which the collector operates *decreases* with increasing temperature. Equations 5.4 and 5.5, which show this clearly, might be expressed in terms of a collector efficiency, η_c, representing the fraction of the incident energy which is extracted from the collector. If we can expect the engine efficiency, η_e, to be half the Carnot efficiency, we shall have

$$\eta_e = \tfrac{1}{2}(1 - T_a/T), \qquad\qquad 6.11$$

when the sink is at atmospheric temperature T_a. Then the combined efficiency of the collector and engine is

$$\eta = \eta_c \times \eta_e. \qquad\qquad 6.12$$

When the rate of incidence of solar energy on the collector is P, the power available from the engine is then

$$P_2 = \eta P. \qquad\qquad 6.13$$

For any given value of P, an increase in collection temperature T brings an increase in η_e but a decrease in η_c. The result is that the overall efficiency η at first increases and then decreases, as shown in Figure 6.12 for various representative conditions. (This figure also shows the very low values of the overall efficiency which result from such a combination). If we imagine a system which could be controlled so that it always operated at the maximum overall efficiency for the current conditions, we would be able to obtain from it as work the output power P_2 shown in Figure 6.13. We see from this that in the tropics near local noon, we could expect to obtain only some 20–40 W of power per square metre of collector area. The total work obtainable per day can be calculated, using the same procedures as in Chapter 5. We find that even at midsummer in the tropics, the work obtained per day is little more than 100 Wh per m². We cannot then expect heat engines with flat-plate collectors to be economic propositions for power production on a large scale, even if the intermittent output could be tolerated, because of the exceptionally large collectors that would be required. Nevertheless, they have been used already for powering small local enterprises such as the pumping of water or the operation of a remote radio station. A well-

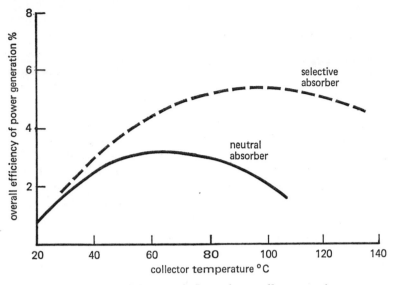

Fig. 6.12 Overall efficiency of flat-plate collector-engine systems
($P = 800$ W/m²)

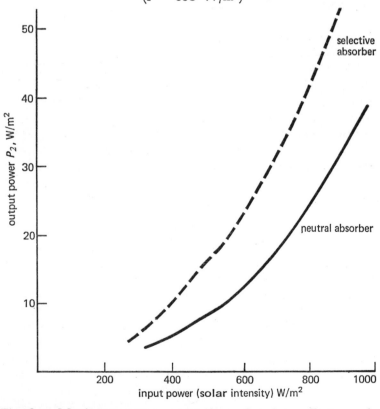

Fig. 6.13 Maximum power output from flat-plate collector-engine systems

known example, which has been put into quantity production, is the 'Somor' device, made in Italy in various sizes with ratings up to 5 kW.

Some engines operating with flat-plate collectors (certain varieties of the 'Somor', for instance) have flat-mirror concentrators attached, to increase the effective intensity of radiation at the absorber plate. This is a step towards more powerful concentrators, which we consider in the next section. The effect of quite modest concentrators is sufficient to warrant the weighing of the advantages against the disadvantage of the need to move the system to follow the sun's apparent motion. Using the methods of this section, we find, for example, that with an effective CR of 2, the maximum power output obtainable from a collector-engine combination with a neutral absorber, is over 80 W/m² when the solar intensity is 800 W/m², (of which 90 W/m² is considered to be diffuse and not amenable to concentration).

6.6 *Heat engines with concentrating collectors*

We saw in Chapter 4 that very high temperatures can be obtained by focussing the sun's radiation by parabolic reflectors. Although such a reflector collects no more energy than a flat one of the same size, the concentration may allow the associated engine to operate with a much higher maximum temperature, and thus, from equation 6.9, a much higher efficiency. Then more of the incident energy will be available as work.

We have just seen that a modest concentration brings a large increase in power output. When the concentration ratio is much more than this, we are justified in taking a somewhat different approach.

We might suppose, for the sake of a simple analysis, that the object upon which the sun's radiation is focussed, which is to form the source of the engine, operates at a temperature at which the bulk of its heat losses to the surroundings occurs by radiation. As in the case of earlier studies, we will assume also that the source is insulated in such a way that it radiates only from the exposed surface. Since the radiating surface will in practice nearly always be greater than that exposed to solar radiation, we must allow for this, as before, by a suitable correction in the definition of the effective CR. Then the heat balance per unit area for the source, under the assumed conditions, would be represented by

$$a_s P \times CR = P_e + \epsilon\sigma\, T^4, \qquad 6.14$$

where CR is the effective concentration ratio of the reflector (see Section 4.5). Thus the rate of extraction of energy per unit area of source is

$$P_e = a_s P \times CR - \epsilon\sigma\, T^4$$

and if we assume, as before, that we can use this source to drive a heat engine having half the Carnot efficiency and using a sink at a temperature of about 290 K (17°C), the power available per unit area of source is

$$P_3 = \tfrac{1}{2}\!\left(1 - \frac{T}{T_a}\right)\!\left(a_s P \times CR - \epsilon\sigma\, T^4\right) \qquad 6.15$$

and the power available per unit area of *collector* would be approximately

$$P_2 = \frac{P_3}{CR} = \tfrac{1}{2}\!\left(1 - \frac{T}{T_a}\right)\!\left(a_s P - \frac{\epsilon\sigma T^4}{CR}\right). \qquad 6.16$$

As in the case of the flat-plate collector and engine, there is a particular value of the source temperature T, for which P_2 is a maximum for any given condition. However, by applying equation 6.15, it is found that this optimum source temperature is for many cases much higher than common metals can tolerate. For any practicable system, we cannot expect to operate with a source temperature higher than about 1000 K (about 700°C). If we set this as an upper limit and calculate the maximum work available at this source temperature, or at the optimum temperature if that is lower, we obtain the curves of Figure 6.14. With this limitation, it is also found that no advantage is to be gained by using an effective concentration ratio greater than about 1000, or by using a selective absorber as can be seen from the figure. Comparing this with Figure 6.13, however, we can see that a very considerable increase in output power is now available, with the maximum overall efficiency amounting to nearly 30%. Minor losses which we have not included, and greater radiation losses than we have assumed here would undoubtedly degrade this considerably in practice, but nevertheless, there is a good possibility of obtaining output powers approaching 200W per m² of collector area in strong tropical sunshine, with a total energy delivery amounting to as much as one kWh per m² per day.

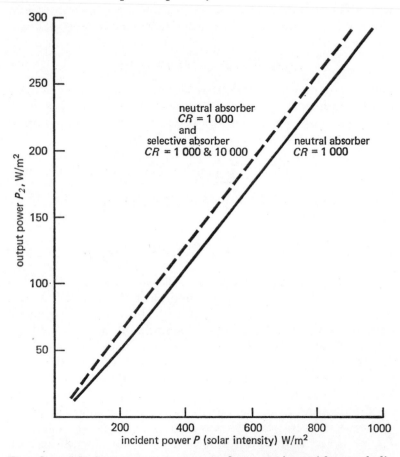

Fig. 6.14 Maximum power output from engine with parabolic concentrator

Against this must be set the rather severe practical limitations on the size of parabolic mirrors and the complications involved in the need for the mirror, or a flat heliostat, to be tracked round to follow the sun. We shall return later to a consideration of these difficulties.

6.7 *Cooling by solar power*

At first it seems a contradiction to state that solar power can be used for cooling. But when the sun's energy is used to drive refrigeration plants, cooling is obtained in just those conditions where it is most needed. As well as improving the health and comfort of people working in hot climates with a consequent increase in their productivity, refrigeration permits the preservation and storage of perishable foodstuffs, which otherwise often have very high wastage rates in tropical countries. It has been said that something like a quarter of the world's cultivated food rots before it can be eaten. In conditions of acute shortage, conserving what one has is as important as trying to obtain more. Modest refrigeration, say storage at 5–10°C, can extend the 'shelf-life' of certain foodstuffs, particularly soft fruit and vegetables, by a large factor. An example, well-known to British readers, is the preservation of bananas in ships, rail containers and permanent stores over periods of weeks and months. In some favourable cases, storage from one harvest to the next is possible.

We saw, in Section 5.5, that radiative cooling alone is

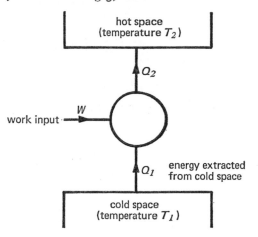

Fig. 6.15 Energy flows in refrigerator

capable of discharging energy at rates of 100–200 W per m² of surface during the night, in very dry climates. This is a very effective method of cooling where meteorological conditions are suitable, and it is questionable whether any other method can compete with it economically. However, we should examine the possibilities of other forms of refrigeration for use in less favourable climates.

The simplest concept of a refrigerator is as a reversed heat engine. Since we have just been dealing with engines, it is convenient to treat refrigerators here.

A reversed engine cycle is simply one similar to those of Figures 6.8 to 6.11, traversed in an anti-clockwise rather than a clockwise direction. As with engines, it is again essential that there be a source and sink, but the direction of the energy flows is reversed. If Figure 6.15 represents a generalised refrigerator, we see that its purpose is the extraction of energy at a rate Q_1 from the cold space, through the expenditure of work at rate W. An essential consequence of this is the rejection of energy in the form of heat into the hot space. (In a domestic refrigerator the hot space is the kitchen, energy being rejected to it from the coils at the back of the machine). The effectiveness of a refrigerator is usually represented by its COEFFICIENT OF PERFORMANCE, which is the ratio of energy extracted from the cold space to the work expended. That is

$$CP = Q_1/W. \qquad 6.17$$

Since energy conservation (or the First Law) requires that $Q_1 + W = Q_2$, we may write

$$CP = Q_1/(Q_2 - Q_1). \qquad 6.18$$

By arguments similar to those used when considering heat engines, it can be shown that no machine can have a CP higher than one operating in a reversed Carnot cycle. For that case, it is found, as for the Carnot engine, that

$$Q_1/T_1 = Q_2/T_2, \qquad 6.19$$

so that the CP can be written

$$CP = T_1/(T_2 - T_1). \qquad 6.20$$

In a typical case we might have a cold-space temperature T_1 of 273 K (0°C) and the hot-space, which in practice would have to be the atmosphere, the earth or a body of water, at a temperature T_2 not more than about 300 K (27°C). For this case, the highest possible CP is about 10. That is, the energy extracted from the cold space would be about ten times the work expended.

In practical refrigerators, the performance is necessarily a good deal worse than this. A representative arrangement is

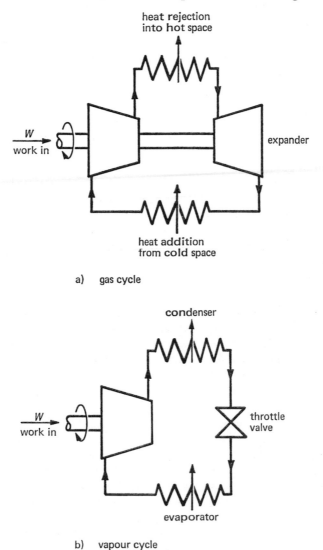

a)   gas cycle

b)   vapour cycle

Fig. 6.16 Simple refrigerators

shown in Figure 6.16(a), in which the compression part of the cycle is shown executed in the pump and the expansion part in a separate expander. Irreversibilities in these processes, friction and heat transfer in various parts of the system seriously degrade the performance. A major factor is that to keep the heat exchangers to a reasonable size, it is necessary for them to operate at a temperature difference of some 5–10 K above or below the temperatures of the hot and cold spaces respectively. The result of these practical restrictions is that refrigerators rarely have CP's greater than 4 under any circumstances. We see from equation 6.20 that the ideal CP increases as the temperatures of the hot and cold spaces are made closer, but in that circumstance, the necessary temperature difference across the heat exchangers becomes a dominating influence and offsets the potential gain.

Hardly any real refrigerators operate with a simple cycle using a gas as the working fluid. Many use a fluid which is at times a liquid and at other times in the cycle a vapour. Fluids commonly used are ammonia (NH_3) and compounds of fluorine such as $C\ Cl_2F_2$. A typical arrangement is shown in Figure 6.16(b). Since the fluid is liquid after leaving the condenser in the hot-space, its pressure is reduced simply by passing it through a throttle valve instead of an expander. The effect of this further irreversibility is a small price to pay for the reduction in mechanical complexity. It hardly needs to be said that the advantages of using a vapour cycle or any other system do not include contravention of the Second Law, so we may continue to think of a CP of 4 as about the highest we can expect from any refrigerator under the circumstances we are considering.

It follows that a good refrigerator, driven by a high-temperature Stirling cycle engine with a parabolic reflector, would permit the extraction of a maximum of about 700–800 W of energy from the cold-space, per m^2 of collector area. If on the other hand, low-temperature, flat-plate collectors were employed, we could expect to extract no more than about 100 W per m^2 of collector area, in strong tropical sunshine.

Although these results are as good as, and sometimes better than, can be obtained with radiative cooling, it is only at the expense of great complexity. One might expect it to be profitable

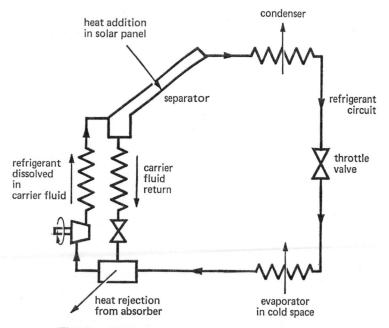

Fig. 6.17 Solar-powered absorption refrigerator

to look for other systems combining the functions of engine and refrigerator and avoiding some of the losses associated with operating them separately. This is, in effect, what happens in the ABSORPTION REFRIGERATOR, illustrated in Figure 6.17, which is well adapted to solar operation, since only a heat input is required. The compressor of the vapour cycle refrigerator is replaced by a system in which the refrigerant, (perhaps ammonia) is dissolved in another (in that case water) at low temperature, pumped to a higher pressure and the fluids separated by heating. The process is possible because the solubility of the refrigerant is much higher at low temperatures than at high temperatures. Since only a liquid is pumped, with negligible compression, the work input is negligible and the system is driven, in effect, by the energy input to the generator. The thermodynamic analysis of such a system is not easy and we need not attempt it here. It is sufficient to note that an additional heat-rejection process occurs when the refrigerant dissolves in the carrier fluid. This is because the mixed state is one of lower energy than the separated state. (Work must be done to separate a mixture, as occurs in the generator). Thus we see that the absorption refrigerator does not eliminate the low-temperature heat-rejection process which takes place in the engine of a combined engine-refrigerator system. It transpires that the overall performance at a given collector temperature is similar in both cases.

It will be instructive to conclude this section by making a rough estimate of the probable cold-storage capacity of a building cooled by systems operating from roof-mounted flat-plate collectors. We shall consider the case of a rectangular building, of length L, width $\frac{1}{3}L$ and height $\frac{1}{5}L$. The external surface area of the walls is thus about $\frac{1}{2}L^2$. We suppose that the whole roof is occupied by collectors which also shade the building so that heat input through the roof is negligible.

The calculations will be simpler if we can assume at the outset an approximate equilibrium temperature for the outer surfaces of the walls. This is determined by heat input by long-wave radiation and some direct and reflected solar radiation (minimised by painting the walls white) and by convective and radiative heat exchange with the surroundings. Evidently, the walls are subject to the same kind of energy exchange processes as other objects in the vicinity, except that heat transfer by conduction to the cooler interior of the building is taking place. We might reasonably suppose then, for a rough calculation, that the surface temperature remains at all times close to the mean of the outside air temperature. In this state, we shall neglect all transfer processes other than the conduction through the wall. The rate of heat transfer through the walls of a building depends upon the difference in air temperature between the inside and the outside, the conductivity of the walls themselves and the nature of the convection in the air on each side. It is usual to combine the effects in a single overall heat transfer coefficient U such that the rate of energy transmission per unit area of wall is given by

$$q = U \Delta T, \qquad 6.21$$

where ΔT is the difference between the inside and outside temperatures. A wide range of values of U can be obtained by the use of insulating materals and different wall designs. A reasonably low value, obtainable without undue expense, would be about 1·0 W/m² K, so that the daily total heat transfer into our building would be about 12 $L^2 \Delta T$ Wh, with L in m and ΔT in kelvins.

By methods developed earlier, we find that it would be reasonable to expect, for a tropical climate with cloudless conditions, a rate of extraction by the refrigeration system of about 500 Wh of energy per day per m² of collector surface. For our rough calculations, we will assume this to be the same, whatever the temperature maintained in the cold-space. If we take the roof area to be $\frac{1}{3}L^2$, an energy balance for one day requires that

$$500. \tfrac{1}{3}L^2 = 12 \ L^2 \ \Delta T,$$

from which we find $\Delta T = 14$ K. Thus the cold space is kept at 14 K below the mean ambient temperature.

A rough comparison can now be made with the results of radiative cooling where conditions are dry enough to use this. The performance of a hypothetical radiative cooling system was shown in Figure 5.9, for an average night-time temperature of 300 K. It is found from this figure that the rate of energy extraction P_c varies with the cold-space temperature T according to the relation

$$P_c = 9\cdot5 \ (T - 275) \qquad\qquad 6.22$$

W per m² of radiator, when T is in kelvins. If it can be operated at this rate for the equivalent of a 6-hour period per day, and we assume a mean ambient temperature of 305 K in using equation 6.21 to obtain the input, we have to solve the equation

$$6. \ 9\cdot5(T - 275).\tfrac{1}{3}L^2 = 12 \ L^2 \ (305 - T).$$

The solution is $T = 289$ K, representing a ΔT of 16 K when radiative cooling is used. Whilst these calculations are of course, very crude, considering the simplicity of this system we can only conclude that refrigeration systems cannot seriously compete with it in situations where both could be used.

At this point we shall leave the consideration of purely mechanical systems. Whilst their possible uses in driving machines for manufacture, pumps for water distribution and so on are potentially very numerous, there are many instances in which it is more desirable to have our power available in the form of electricity. In the next two chapters, therefore, we shall turn to an examination of the possibilities of the production of electricity by solar-powered devices other than generators driven by heat engines.

7

CONVERSION OF SOLAR ENERGY INTO ELECTRICITY

Electricity is of two kinds, positive and negative. The difference is, I presume, that one comes a little more expensive, but is more durable, the other is a cheaper thing, but the moths get into it.

STEPHEN LEACOCK (1869–1944)

THERE is no doubt that the most convenient way of providing power is by electricity. Most of the world's electricity is generated by rotating electromagnetic machines, driven by heat engines of one kind or another. We have seen in the preceding chapter that, except where concentrators are used, solar power units involving heat engines are very inefficient. This is because of the fundamental thermodynamic limitations fixed by the maximum and minimum temperatures of the engine cycle. When concentrators are employed, it is possible to obtain mechanical power at rates of the order of 100 W per m² of collector area, but the complexity of the system makes this a very unpromising method for driving electrical generators other than small units, with outputs of a few tens or hundreds of watts. In this chapter, we shall examine other ways of obtaining electrical power by using the sun's energy, to see if it is possible to improve substantially on these figures. To make a good job of this, we would really have to find a means of evading the constraints of the Second Law of Thermodynamics. As we shall see, this is not simply a matter of replacing the mechanical engine by some other device.

Interchange between electrical and mechanical forms of energy can proceed in principle without loss, as in, say, an ideal solenoid or motor. Electrical machines for converting one form of energy to another are normally very efficient, the efficiency sometimes exceeding 90% in the case of large machines (in which the losses are proportionally lower than in smaller machines). It is, therefore, not unreasonable to think of a perfect machine, with 100% efficiency, and thus capable of working reversibly. But such a machine is not different in principle from an ideal reversible mechanical engine. Then we must expect the 2nd Law to apply to any system for using it. As before, it is in the conversion of energy in the form of heat into work in a continuous manner that the limitations occur. We must conclude that in using solar energy, we could only avoid these limitations if our machine is one in which we do not need an initial stage, involving the conversion of the radiation into heat. Though we must enquire into this possibility we should still look at more conventional systems which might be capable of working closer to the thermodynamic limitations than can the mechanical systems, or of giving as good a performance more simply or economically. In this chapter we shall first consider some principles and then examine some devices for producing electricity which do have an initial heating stage. The advantages and disadvantages of potentially more exciting devices which operate without heating stages are discussed in the succeeding chapter.

7.1 Electrical current, potential and power

The conventional view of the flow of electricity through a circuit is of the passage of a CURRENT under the influence of a POTENTIAL DIFFERENCE. This view is often called to mind by analogies involving the flow of water through pipes connecting tanks at different levels. The foundations of the subject were laid before it was realised that an electric current consists of the motion of electrons and early conventions are the opposite of those which would be chosen now from an electronic point of view. Since, in this chapter, we shall be dealing entirely with the behaviour of electrons, it might be useful at this point to introduce another review section so that the possibility of confusion between the conventional and electronic viewpoints is lessened.

The allocation of the sign (positive or negative) to a quantity is at first arbitrary. In the case of electricity, the initial allocation was made by FRANKLIN (1747), when deciding that the electric charge acquired by a glass rod, when rubbed with silk, would be given a *positive* sign. A vast body of knowledge of electrostatic behaviour was built up over the succeeding century or so. We need not concern ourselves with this here, except to remark that to this period belongs COULOMB's establishment by experiment of the important law which bears his name, giving the electrostatic force experienced by two small charged bodies (1785). If the charges on the two objects are q_1 and q_2 and the distance between them is x, this force is found to be proportional to q_1q_2/x^2, being an attraction if the charges are of opposite sign, a repulsion if of the same sign. Using this result, we can define the unit charge as that residing on each of two equally-charged small bodies placed at some agreed distance apart and experiencing some agreed force. A unit of charge defined in this way is appropriately called the COULOMB.

A more fundamental unit of charge might now be chosen to be equal to the most certainly reproducible charge known, the charge on the electron. The existence of the electron was first revealed in the 'cathode rays' emitted from the negative electrode, or cathode, of a gas discharge tube. By making measurements of the energy of these rays and the manner in which they were deflected by electrostatic forces, J. J. THOMSON (1897) showed that they could be considered to consist of minute particles, each having a charge equivalent to about 1.6×10^{-19} coulomb, and negative in sign by Franklin's convention. (The name electron is derived from the Greek word for amber, on which a negative charge can be induced by rubbing). We can now think of the occurrence of negative charge anywhere as arising from the presence of electrons there, in excess of those already present in (electrically-neutral) atoms. The origins of electrons are atoms and molecules, so that if electrons have passed to a body and given it a negative charge, they have left behind them atoms or molecules deficient in electrons. These are the IONS, which evidently have a positive charge. Thus, when Franklin's glass rod was rubbed with silk, the actual effect was the transfer of electrons to the silk, leaving behind positive ions in the glass.

Now if an electron is to be separated from its parent ion, work will have to be done against the Coulomb force which attracts them together. This is rather like stretching a spring joining them (though it would be an unusual spring, capable of stretching indefinitely and getting weaker the further it is stretched). We saw in Chapter 3 that work done on a system like this is often conveniently represented by an increase of the energy of the system. This energy has to come from somewhere. One way of providing it is by using a battery, in which a chemical reaction, going from a state of higher energy to one of lower energy, provides the work necessary to separate large numbers of electrons from their ions.

There are many other ways of achieving charge separation, some of which we will examine in detail later in this chapter. Any of these could be used to bring about the situation shown in Figure 7.1a, in which two objects, separated by a vacant space, have their charge distribution altered by a device capable of charge separation. We have just seen that in such a process, work is done against Coulomb forces. These forces act so as to recombine the charges. Thus, in Figure 7.1a, if an electron finds itself free of the CATHODE (by a process to be considered in Section 7.5), it will be urged by Coulomb forces towards the ANODE. These forces will accelerate it until, just as it reaches the anode, it has acquired kinetic energy equal to the work done to separate it from the ion originally. Upon impact, this energy will be communicated to the 'fixed' ions of the anode whose increased vibration we interpret as a rise in temperature. The work put in during charge separation is thus ultimately converted into internal energy. A somewhat different situation arises if we connect the anode and cathode by an electrical conductor, as in Figure 7.1b. The electron is still urged towards the anode, but its way is impeded by the fixed ions of the conductor. It pursues a very devious route, scattered by frequent collisions with ions. Instead of accelerating freely, it moves along this route with an average speed of some hundred of metres per second, but progresses towards the anode only at a very low rate, usually a few centimetres per second. Between interactions, the Coulomb forces increase the component of the electron's velocity in the direction of the anode, so that it acquires some kinetic energy. But at each interaction, it gives up, on the average, as much as it acquires, on the average, between them. The ions are thereby set vibrating more vigorously; that is, the temperature of the conductor is increased all the way along. Again the work of the original charge separation is converted into internal energy. This is a simplified view of the phenomenon of RESIST- ANCE to the passage of current in conductors. The internal

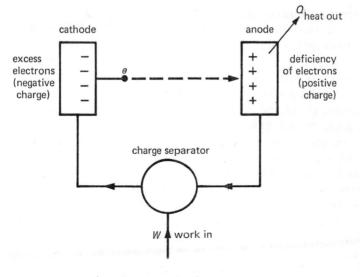

a)   electron motion in vacuum

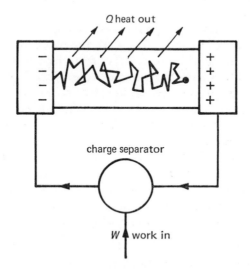

b)   electron motion in conductor

Fig. 7.1 Charge separation and electron motion

SOLAR ENERGY FOR MAN

energy acquired by the ions is known as JOULE HEAT.
It is customary to speak in these situations of the existence of
a POTENTIAL DIFFERENCE between cathode and anode. By
definition, the potential difference V, between two points A and
B, is equal to the work that would have to be done to move a test
body, having a small positive charge, from A to B, divided by
the charge carried. That is, V is the work done per unit positive
charge. In our examples, it is evident that with a positively-
charged test body instead of the electron, work would have to be
done to move it from cathode to anode. Then, by definition, the
anode is at a higher POTENTIAL than the cathode. The anode is
therefore called the positive electrode, and we see that this is
consistent with the sign of the excess charges there. But, before it
was known that the actual carriers of charge were negatively
charged, it was assumed that the current flowed from the
positive to the negative electrode, that is, in the opposite
direction to the real electron flow. Even today, it is often considered to be convenient to represent current this way. We then
speak of the CONVENTIONAL CURRENT rather than of the
electron current.

The relationship between the quantities potential difference,
current, resistance and power can be established quite easily
with the simple model we have been using. If the potential
difference across the conductor is V, the work to be done in
moving an electron with charge $(-e)$ from cathode to anode is
$-eV$. The negative sign here means that work is done by the
system; we need not put any in, but we have to provide something (the ions) to take up the energy. When N electrons move
per unit time, the rate of doing work, or the POWER, is $-NeV$.
But $-Ne$ is the charge moved per unit time, which is, by definition, the CURRENT I. Thus the power used is IV. Experiments
show that for many conductors, the current I is proportional to
the potential difference V, so that $I = $ constant $\times V$. The constant is called the conductance, though we are more accustomed
to using its reciprocal, the RESISTANCE, R. We then have

$$I = V/R, \qquad\qquad 7.1$$

the familiar relation known as OHM'S LAW (1827). The power
equation then takes the alternative forms

$$P = IV = I^2R = V^2/R. \qquad\qquad 7.2$$

Following these definitions, we can now turn to a more detailed examination of the behaviour of electrons in solids of various kinds, building further upon our model of the solid, with its framework of ions, more-or-less fixed relative to each other but perpetually oscillating about their mean positions.

7.2 *Electrons in solids*

We saw in Chapter 3 how the concept of energy levels has arisen in the case of the single atom. It is possible to calculate the magnitude of the energy level associated with each state that an electron could occupy, when it belongs to an isolated atom. In a few cases, the energy levels have been calculated for a combination of two or three atoms. When a very large number of atoms or molecules are gathered together in a solid, however, it is no longer practicable to calculate the actual values of the energy levels because they are so numerous. A particular electron is influenced by all the nuclei in its vicinity and by other electrons, so that it has a multitude of energy levels open to it. Some of these levels are very close together and it can happen that there are numbers of possible states having virtually the same energy. It is convenient, then, to think of ENERGY BANDS rather than of separate levels, each band consisting of very many levels with energies of similar magnitude. Because the bands are made up of permitted levels, however, we must not forget that quantum rules apply to movements of electrons from level to level within a band. Moreover, energies lying outside a band, that is, in a region where there are no permitted levels, are inaccessible to the electron. Electrons moving between levels in a band can only do so by acquiring or losing energy, however small the amount may be. We shall have to consider later how this can come about, but for the moment we can see that electrons may acquire energy by interactions with other electrons or with the oscillating ions which make up the regular structure of the solid. An electron is, nevertheless, not entirely free to move from level to level since nature seems to operate the restriction that no two electrons may occupy a given level. In more detailed studies, this limitation is encountered in what is called the EXCLUSION PRINCIPLE due to PAULI (1925). The idea is rendered plausible by noting that if two electrons did have exactly the same state, it would not be possible to distinguish between them. They would then appear in some circumstances to be a single particle having twice the mass and charge of an electron, something which has never been observed.

We can now relate the concept of energy bands to certain properties of various substances in bulk. In particular, we can see why they vary in their capacity to conduct electricity, that is, in the mobility of their electrons. Figure 7.2 shows diagrammatically the highest two energy bands of three substances in the solid state. These top two bands are called the VALENCE BAND and the CONDUCTION BAND, according to the occupancy of their energy levels by electrons, as we shall see later. Between the highest level of the lower band and the lowest level of the upper band there is shown a space, the FORBIDDEN BAND. There are no levels in this space and no electron can have a value of energy lying in it. The width of such a space, called the ENERGY GAP, differs for different materials.

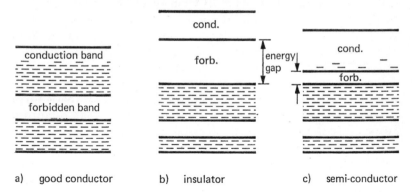

a)   good conductor   b)   insulator   c)   semi-conductor

Fig. 7.2 Representation of energy band structure in solids

In part (a) of the diagram, the nature of the substance is such that the energy levels of one permitted band are all occupied, but that those of the next highest permitted band are only partly occupied. Above the highest filled energy level there are many more levels differing only slightly in energy from it. The electrons in this band, the CONDUCTION ELECTRONS, will not now be confined to a given state and can move from one level to another, since this requires only trifling changes in energy. These are the characteristics of a good CONDUCTOR, such as copper or silver. Parts (b) and (c) of Figure 7.2 show substances in which the highest occupied band is full. Electrons in this band, the VALENCE ELECTRONS, are confined to their present energy levels, since there are no other levels to which they can readily move. Substances with these characteristics are not good conductors. The only way in which electrons can become mobile is to acquire enough energy to cross the forbidden band into the next band in which there are many levels available to them. Whether or not they can do this depends upon the magnitude of the energy gap and the ways in which electrons can obtain energy.

We saw, previously, that one way is by the absorption of radiation. Where this occurs, a substance which is normally an INSULATOR becomes a conductor when suitable radiation falls on it. Later, we shall look at this process again. Electrons can also acquire energy from the positive nuclei via the electrostatic forces between them. These nuclei, together with the electrons in the lower energy levels which cannot escape from them, are the ions. Since they are atoms or molecules which have lost one or more electrons, they have a nett positive charge. The ions form the regular structure of the lattice of the solid, though as we saw earlier, they are not fixed, but oscillate about their mean positions. As the temperature of the solid rises, these oscillations become more vigorous, and this vigour is communicated to the electrons also. At a sufficiently high temperature, a significant number of electrons have acquired enough energy by this means to cross the gap into the conduction band and the substance changes from being an insulator to being a conductor. When the energy gap is large, this change will only occur at high temperatures. All those substances which we normally think of as insulators—glass, for example,—exhibit this breakdown when the temperature is high enough.

In some substances, the energy gap is narrow enough for a few electrons to be able to cross it at ordinary temperatures. If the temperature is raised, more electrons can cross, so that we have a substance whose conductivity rises with increasing temperature in familiar circumstances. Those substances whose conductivity is low—perhaps a thousandth or a millionth of that of the good conductors—at ordinary temperatures, but rises steeply with increasing temperatures, are called INTRINSIC SEMICON-DUCTORS. Germanium and silicon are examples of elements with these properties. The reader will be aware that much of the recent enormous advance of electronic engineering has depended upon the exploitation of the properties of semiconductors such as these, and of other kinds, which we shall consider shortly.

7.3 *Energy distribution of electrons*

The electrons in the conduction band are able to move freely around in a solid and it is these which can be made to provide a current. Their energies will be constantly changing by small amounts through interaction with each other, and with the oscillating ions. At any instant, however, there will be a nearly steady distribution of electrons among the energy levels, representing the equilibrium state which results from the continual interchange of electrons from one level to another. The distribution of electrons among the highest energy levels is of great importance in certain phenomena which we need to understand, so we shall have to examine this point. The actual evaluation of the distribution is a rather complicated statistical exercise and we cannot attempt it here. However, the reader will see that such an evaluation is possible if we first consider a simpler, but somewhat analogous case, the distribution of the combined reading of a pair of dice. Eleven values of the combined reading, from 2 to 12, are evidently possible. But if the dice are cast many times, some readings will be commoner than others. The commonest, for example, will be 7. This is because that number can be obtained by more combinations than for any other number (7 can be made by 1 + 6, 2 + 5, 3 + 4, 4 + 3, 5 + 2, and 6 + 1, whereas 3 can be made only by 1 + 2 and 2 + 1). If we cast the dice enough times and count the number of occasions on which each of the possible totals occur, we obtain a distribution of readings among the permitted levels. Now we can easily predict what this distribution will be like. The reader can verify that the readings of the dice can combine in a total of 36 different ways. We have just seen that 6 of these result in a reading of 7 and two of them result in 3. Then we can show, as in Figure 7.3, the fraction of the total number of throws likely to result in each of the possible values (6/36 for 7, 2/36 for 3 and so on). Now, if we could imagine a huge number of pairs of dice being thrown repeatedly, the distribution of combined readings among them at any instant would be according to this

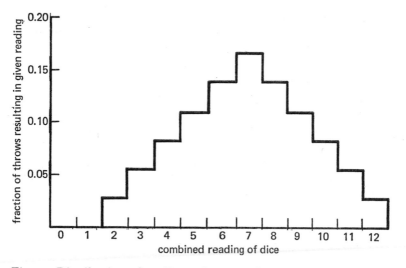

Fig. 7.3 Distribution of readings of a pair of dice, thrown repeatedly

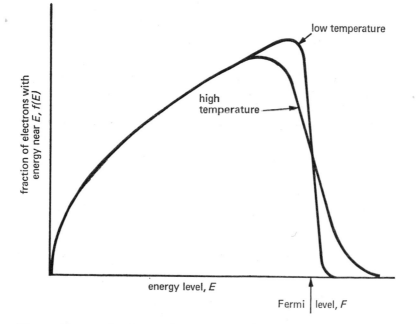

Fig. 7.4 Energy distribution in conductor (Fermi-Dirac distribution)

pattern. It would be the same at any other instant. The readings of a particular pair of dice would, however, be constantly changing. We can, similarly, imagine the distribution of electrons among the available energy levels, though the constraints on that particular distribution are very different.

For good electrical conductors, the appropriate distribution is that devised by FERMI and DIRAC in 1926, and therefore known as the FERMI-DIRAC DISTRIBUTION. It may be presented as in Figure 7.4. The number of separate energy levels is so great that we do not now have a noticeably step-wise distribution as for the dice; the steps are so small that a continuous curve results. In Figure 7.4, the vertical displacement represents the fraction of electrons having energies within a narrow band in the vicinity of the corresponding value of E. The actual equation of this curve is of the form

$$f(E) = \text{constant} \times \frac{E^{\frac{1}{2}}}{1 + \exp\left[(E - F)/kT\right]}, \qquad 7.3$$

where kT is, as given originally in equation 3.4, about the average kinetic energy of a particle at temperature T. The reader who is not familiar with the properties of the exponential function in the denominator of equation 7.3, will be able to see its effect from the diagram. Except within a few kTs of the energy F, called the FERMI LEVEL, the distribution is unaffected by temperature. But near F, the electron concentration falls rapidly. The effect of increasing temperature is really quite small, but as the temperature rises, more electrons are found with energies above the Fermi level and fewer with energies below it. The smallness of the affect of temperature on the energies of the electrons can be visualised if we note that for metals at ordinary temperatures, kT is typically about $1/200$ of F and at $T = 1\,000$ K, kT is still only about $1/70$ of F.

We now have the essential information to enable us to discuss the behaviour of the first device for the conversion of heat to electricity. Before doing so, it will be useful to introduce at this point the ELECTRON-VOLT as a unit of energy. In processes on an atomic scale, conventional units such as the joule are inconveniently large. For instance, the average kinetic energy of an electron at ordinary temperatures is about 4×10^{-21} J. The electron-volt is a much smaller unit. It is equivalent to the work which would have to be done to move an electron, with its negative charge, between any two points whose electrical potential differs by 1 volt. Evidently, it is an appropriate unit for the events we are discussing. The relation between the joule and the electron-volt is readily obtained. The joule is the energy transferred when work is done at a rate of 1 watt for a period of 1 second. An instance in which work at a rate of 1 watt is being done is when a current of 1 amp passes through a conductor across which the potential difference is 1 volt. For the current to be 1A, charge must be passing at a rate of 1 coulomb per second, so that 6.25×10^{18} electrons must be passing per second. Thus, $1J = 1 \ W{-}s = 1A{-}V{-}s = 6.25 \times 10^{18}$ $\frac{e}{s}{-}V{-}s = 6.25 \times 10^{18}$ eV, or $1eV = 1.6 \times 10^{-19}J$.

The constant k in our equations was first introduced in our discussion of kinetic theory in Chapter 3. It is called BOLTZ-MANN'S CONSTANT, and it has a value in terms of the electron-volt of $1/11\,600$ eV per kelvin. Consequently, at ordinary temperatures (say $T = 300$ K) kT is about 0·026 eV and at $T = 1\,000$ K, kT is about 0·086 eV. By comparison, Fermi levels are usually in the range from 5 to 10 eV.

With the above outline of electronic energy levels and a convenient unit to represent them in, we can now consider the first type of device for the direct generation of electricity, known as the THERMIONIC GENERATOR. This kind of device has been extensively studied during the last few decades, and is now expected to play an important role in electricity generation in the future. It is, however, still not well known and we might usefully examine it in some detail.

7.4 The thermionic generator

Figure 7.5 illustrates the thermionic generator in principle. It is based on an effect, known as THERMIONIC EMISSION, noticed by EDISON in 1883, though the first suggestion for such a generator is said to have been made by SCHLICHTER in 1915. When one of the electrodes, later to become the CATHODE, is heated to a high enough temperature, a significant proportion of its electrons acquire enough energy to escape altogether from the surface. We shall look a little closer at this process later, but for the moment need only note that whatever energy is required for this, some electrons will have it if the temperature is high

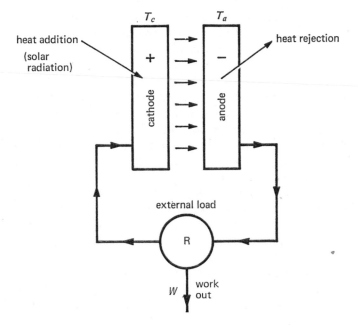

Fig. 7.5 Thermionic generator

enough. If there is another electrode, the ANODE, near by, the emitted electrons can be encouraged to collect on it. This would be a transient occurrence if only the two electrodes were involved because the growing negative charge on the anode would repel the electrons emitted subsequently and eventually no more would be able to reach it. However, in a thermionic generator, the anode is connected through an external circuit back to the cathode. Then a stream of electrons, constituting a current, passes through this circuit, doing work in it. In Figure 7.5, the external load is represented by the resistor R, but in practice would be any device that had to be driven, such as an electric motor. The thermionic generator thus functions by using some of the energy used to heat the cathode (in the case we are interested in, this energy is provided in the form of solar radiation) to drive electrons through the external load, doing useful work there.

We can see at once that this conversion of solar energy into work is not achieved without loss, so that the question of the efficiency of the device arises. Electrons will only be able to leave the cathode if it is made hot, and then we shall lose energy by radiation to the surroundings. Some of this radiation will fall on the anode, which, if it is allowed to warm up too much, will also emit electrons. If any of these pass across to the cathode, they would constitute a reduction in the nett current. To keep the anode cool, it too must be made to lose energy by radiation or some other cooling mechanism. In fact, because the thermionic generator is obliged to reject energy to a cooler sink in this way, we see that it is restricted in the same manner as any other heat engine and we must expect its efficiency to be similarly limited. Before determining the efficiency of such a device, however, we must first examine the process of thermionic emission of electrons from a hot solid.

7.5 *Thermionic emission*

Although it appears that large numbers of electrons are present in conductors in conditions of almost complete freedom, it is evident from common experience that they do not readily escape from the surface. The reason for this is not hard to find if we confine our attention at first to a single electron in an isolated piece of metal. As long as it remains in the solid, the surrounding material has no nett electric charge, since there are equal numbers of electrons and ions present in it. If the electron is detached from the solid, however, it leaves behind a deficiency of negative charge and hence an excess of positive charge. This excess positive charge may not be actually located at a particular point in the solid, but it is possible to show that for any given position of the electron outside the solid, the distribution of charges inside it is *as if* a positive charge were located at a particular point. This point is the IMAGE POINT, shown in Figure 7.6.

The nett attraction on the electron by the solid surface is then as given by the Coulomb equation for the attraction between a pair of positive and negative charges, separated by a distance $2x$,

$$f = \text{constant} \times \frac{e^2}{(2x)^2}. \qquad 7.4$$

The exact value of x cannot be inserted into this equation, because on an atomic scale, the surface of the solid is not smooth as shown in the diagram, and we cannot say with certainty where the virtual image is situated. However, we can see that work would have to be done to move the electron away from the surface against this progressively-weakening force. The work

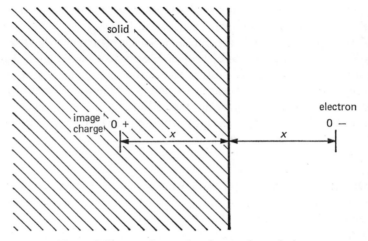

Fig. 7.6 Image charge in electronic emission

necessary to move the electron to an infinite distance away is called the WORK FUNCTION, φ. It can be calculated approximately on the basis of equation 7.4. Because the retarding force falls so rapidly with distance, it is found that very little more work is necessary to move the electron after it has been separated from the surface by more than a few micrometres. Thus, we can say to a fair approximation, that the kinetic energy which an electron inside the solid would have to acquire to enable it to move to just outside the solid is equal to the work function. For metals, the work function has values from about 2 to 5 eV. Since this is very large compared with kT at ordinary temperatures, our discussion in the previous section would lead us to expect that virtually no electrons could escape from metal surfaces at ordinary temperatures, and that is usually the case.

We saw that those electrons which are affected by a rising temperature have energies near to the Fermi level. Accordingly, we can think of φ as the further energy required to move an electron from the Fermi level F to a position outside the solid.

From the distribution of energy among the electrons, given by the Fermi-Dirac expression, equation 7.3, it is shown that the number of electrons with energy equal to, or greater than $(F + \varphi)$ is proportional to $T^2 \exp(-\varphi/kT)$. Then at any temperature T, the number of electrons escaping from unit area of the surface per unit time is given by

$$j = AT^2 \exp(-\varphi/kT), \qquad 7.5$$

where A is a constant. Since a stream of electrons is a current, j is called the CURRENT DENSITY. Equation 7.5. is known as the RICHARDSON EQUATION (The first studies of thermionic emission by RICHARDSON date from 1903). A is a universal constant having a value of about $1\cdot2 \times 10^6$ amp per m² per K². It is apparent that the process of electron emission from solids is very similar to the evaporation of molecules from liquids, which we considered in Section 5.6 without so much detail. In evaporation also, work has to be done to separate the molecule from its fellows in the bulk liquid and the proportion of molecules possessing enough energy for this rises rapidly with temperature. We can now see how the emitted current of electrons, on which the thermionic generator depends, varies with temperature. Table 7.1. shows the current density at various temperatures for typical electrodes of bare tungsten, caesium-coated tungsten, and caesium-coated silver oxide, whose work functions are about 4·5, 2 and 1 eV respectively:

TABLE 7.1

Thermionic current density (A/m^2) at various temperatures

	500 K	1 000 K	1 500 K	2 000 K	2 500 K
Tungsten, W			0·1	25	$6\cdot5 \times 10^3$
Caesium on Tungsten, Cs–W		100	5×10^6	4×10^7	
Caesium on Silver Oxide, Cs–Ag	25	1×10^7			

This table shows how strongly the emission depends on the work function at a given temperature and that high temperatures are required to obtain useful currents (of the order of 10 A per cm², or 10^5 A per m²). In practice, there are two important limitations on the temperature which may be used. Firstly, when the current density becomes too great, the internal resistance loss, or Joule loss becomes unacceptable. Secondly, when the temperature is high enough the positive ions themselves attain sufficient energy to escape from their lattice positions and there is a loss of material by evaporation. Other factors affecting the choice of electrode materials will be discussed later.

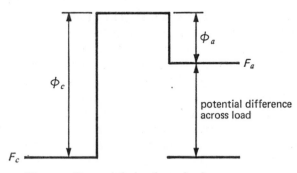

Fig. 7.7 Potentials in thermionic generator

7.6 Output and efficiency of thermionic generators

The variation of potential in a thermionic generator is approximately as shown in Figure 7.7. Electrons are raised from the Fermi level F_c of the cathode by heating it until enough surmount the barrier, represented by the work function φ_c of the cathode, and pass across to the anode. There they come into contact with a cool material and fall to its Fermi level F_a, the excess energy represented by the work function of the anode, φ_a, having to be radiated away to the surroundings. If we arrange that φ_a is less than φ_c, the difference in energy between anode and cathode is available to drive the electrons through the external load and do useful work there.

For this system to work, as described, we need a low φ_a and this means that unless the temperature T_a of the anode is kept low, there will be a significant number of electrons emitted from it and passing in the reverse direction. With a low-temperature anode, the analysis of the power output and efficiency is simplified, since we can then neglect both the back-emission of electrons and the radiation of heat from it, by comparison with the cathode. In what follows, then, we shall suppose that a suitably low temperature can be maintained at the anode. To obtain the kind of temperatures at which thermionic emission is strong, it is obvious that when the heat input to the cathode comes from the sun, it must be provided via a concentrator. When an equilibrium has been reached, the energy input by radiation from the sun, at a rate $a_s P \times CR$ per unit cathode area, will be balanced by the rate of energy loss per unit area, comprising the radiation emission, $\epsilon\sigma T_c^4$ and the electron emission. If the electron current density is j_c, the rate at which energy is needed by electrons to escape is $j_c\varphi_c$. However, because of the distribution of electron energies at a given temperature, which we recently discussed, many of them have energy in excess of the minimum (φ_c) necessary to escape. This small, but significant excess in energy is generally reckoned to be equivalent to an average addition of $2kT_c$. Then the power balance for the cathode is given by

$$a_s P \times CR = j_c(\varphi_c + 2kT_c) + \epsilon\sigma T_c^4. \qquad 7.6$$

We should note here that, as before, the effective concentration ratio CR used in this equation can be defined in such a way as to include the effect of differences between the radiating surface area and the area actually exposed to solar radiation. In a practical thermionic device, the exposed area might be typically about half the total cathode area, with a consequent reduction of attainable effective CR to perhaps 1 000 or less.

If the relation between j_c and T_c, equation 7.5, is now employed, we have

$$a_s P \times CR = (\varphi_c + 2kT_c)AT_c^2 \exp(-\varphi_c/kT_c) + \epsilon\sigma T_c^4,$$
$$7.7$$

so that we find a direct relation between $a_s P \times CR$, the power input per unit emitting area, and the equilibrium value of the cathode temperature T_c. Now if the effective CR is 1 000 or so, then even for strong tropical sunshine, $a_s P \times CR$ will not exceed a million watts per square metre. A surface at a temperature of just over 2 000 K can emit energy at this rate by thermal radiation alone. Thus, we must choose for our cathode material one which emits electrons strongly at temperatures less than this. For that to happen, we could expect from the data of Table 7.1 that the cathode work function must be less than about 3 eV. One of the most extensively investigated devices is the tungsten cathode, impregnated or coated with the alkali metal caesium. The work function is only slightly greater than that of bulk caesium, perhaps 2 eV. Using equation 7.7, and assuming an effective emissivity of about 0·4, which would be typical at the high temperatures concerned, we then find the kind of relation between solar intensity and cathode temperature shown in Figure 7.8. It can be seen that, because the total energy emission changes so rapidly with temperature, the cathode temperature is fairly insensitive to solar intensity.

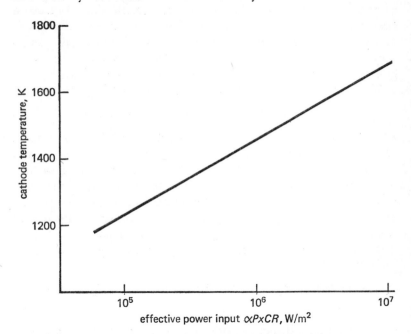

Fig. 7.8 Cathode temperature in thermionic generators

Now the current passing round the circuit per unit area of cathode is the current density j_c. The energy carried by the electrons leaving the cathode is partly given up at the anode, being radiated away or otherwise extracted to keep the anode cool. The remaining energy is represented by the potential drop $(\varphi_c - \varphi_a)$ across the external load. Then the output power, or rate of doing work in the external circuit is, by equation 7.2,

$$P_3 = j_c(\varphi_c - \varphi_a) \qquad 7.8$$

per unit area of cathode. Finally, for comparison, we shall want to obtain the output power per unit area of collector, given by

$$P_2 = P_3/\mathrm{CR}. \qquad\qquad 7.9$$

A suitable anode for use with our supposed Cs—W cathode might be one consisting of a base of silver oxide with a coating of caesium. At operating temperature, the work function φ_a of this type of anode would be about 1 eV.

Using the equilibrium temperatures from Figure 7.8, equation 7.5 to obtain j_c and equations 7.8 and 7.9 to give P_2, we obtain for our example the output powers shown in Figure 7.9 for several values of the effective CR. In arriving at these, it has been assumed that an absorptivity of 0·8 for solar radiation can be obtained at the high cathode temperatures involved.

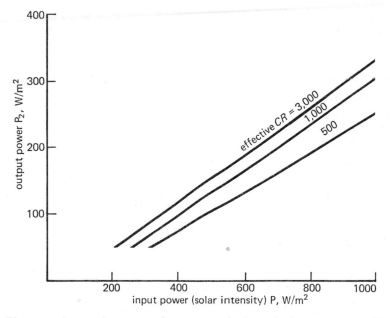

Fig. 7.9 Approximate performance of thermionic generator with concentrator

Although we have used a very simplified analysis in arriving at the typical results of Figure 7.9, we could expect to obtain output powers approaching those shown in a fully developed practical system. One of the difficulties facing the designer of thermionic engines has, however, so far been ignored. This is the SPACE CHARGE effect which arises when we try to use these devices at high current densities. There are then large numbers of electrons in the space between the cathode and the anode and, as they are all similarly charged, they repel each other by Coulomb forces. This repulsion is equivalent to an additional barrier to the escape of electrons from the cathode and it is customary to speak of the SPACE-CHARGE BARRIER in this context. One method commonly used to reduce this difficulty is to arrange that the space between the electrodes is very small, so that not too many electrons are in it at any time. This creates great production difficulties, however, because to be effective, it requires a spacing of the order of a few μm—though this has been done. Another method is to arrange for positive ions to be present in the inter-electrode space to neutralize the charge of the electron cloud. This is usually done by introducing the vapour of caesium into the space. On contact with the hot cathode surface, the caesium atoms are given enough energy to IONISE, that is, for one electron to be freed from the rest of each atom, leaving a positively charged ion behind. These ions are very massive compared with the electrons and are swept by them towards the anode only at a very small rate. Caesium is used for this purpose because the energy required to remove an electron from a caesium atom is smaller than for any other element. This energy is called the FIRST IONIZATION POTEN-TIAL and is about 3·9 eV for caesium. One reason for choosing caesium-loaded electrodes in our example, was their compatibility with caesium vapour; in the case of the tungsten-caesium cathode, this electrode itself can also be arranged to be the source of caesium ions for the system.

One practical disadvantage of the thermionic generator as a primary power source is clear from the potential variation shown in Figure 7.7. There, the potential difference across the external load is shown to be ($\varphi_c - \varphi_a$); in our example, this will be equivalent to 1 volt. So low a potential difference means that the current must be very large for useful values of the output power. For example, if the generator produces an output power of 1 kW, the current in the external circuit would have to be 1 000 A. Such large currents are a great nuisance in practice. To transmit them over any distance requires very thick conductors if the Joule heating is not to represent an unacceptable loss. Moreover, they set up very intense magnetic fields whose inter-actions cause large forces to be exerted between adjacent con-ductors. In practice, then, it is customary to use numbers of generators of this kind connected in series so that their com-bined output voltage is large. Converters and transformers further ensure that the output is provided at much higher potential differences and as an alternating current. From that point on, all our operating experience with devices designed for 110–240V, 50–60 cycle supplies is then directly applicable.

The reader might have noticed a further difficulty which was by-passed in our simplified analysis. It was assumed that the potential drop across the external load was maintained at $(\varphi_c - \varphi_a)$ at all times. It is evident that unless the generator is operating in steady conditions, the resistance R of the external load could not be constant atthe same time. For the potential drop across R will be IR, so that $R = (\varphi_c - \varphi_a)/I$, which will vary with the current I. An arrangement in which R is so ad-justed that the potential drop across it is always $(\varphi_c - \varphi_a)$, as assumed, is called the MATCHED LOAD condition. It is possible to obtain this condition with a continuously-variable external load, but only at the expense of further complexity. Of course, the simplest kind of load with these characteristics is an ordinary storage battery, which can accept a varying current whilst maintaining an almost-constant potential difference between its terminals, determined by the chemical reaction. Another possi-bility is to cause the generator to work at almost constant condi-tions, in spite of variations in the thermal input. For this, a constant-temperature stage is introduced between the input surface and the cathode itself. A few substances with melting-points at suitable temperatures for operation in thermal storage systems of this kind have recently been investigated for use in space vehicles.

Though it would be wrong to minimise the practical diffi-culties in suitably utilising the output power, we shall not want to obscure the principles with too many details. Enough has been done to show how thermionic generators, though much simpler, are able to produce outputs at least as high as those obtained with the conventional kinds of heat engines discussed in the previous chapter. We might usefully enquire into the ultimate limitations on the efficiency of thermionic systems, before passing on to others.

7.7 Thermionic generators and the Carnot efficiency

It was pointed out earlier that thermionic generators, through their rejection of some of the supplied energy to a cooler sink, are subject to the same thermodynamic limitations as other heat engines. As long as the anode temperature is just low enough for its back-emission to be negligible, it is immaterial to the performance of the generator what that temperature actually is. In consequence, it can be quite high, and it is possible to make use of the energy liberated in the anode before it finally reaches its ultimate sink at ordinary temperatures.

We have first to decide what is the highest acceptable anode temperature. An idea can be gained from a consideration of the hypothetical perfect generator. We saw in chapter 6 that no engine can be more efficient than one which is reversible. In that case the relation

$$Q_1/T_1 = Q_2/T_2, \qquad 7.10$$

first encountered as equation 6.10, must be satisfied. In the case of the thermionic generator, reversibility would be possible only if radiation losses were negligible. In such a (hypothetical) situation, the energy Q_1 supplied to the cathode would just be equal to that leaving it with the electrons. The smallest value this can have is $j_c\varphi_c$ per unit area. Similarly, Q_2 is that part of the energy rejected as heat by the anode, which is $j_c\varphi_a$. Then equation 7.10 requires that

$$j_c\varphi_c/T_c = j_c\varphi_a/T_a$$

or
$$T_a = \frac{\varphi_a}{\varphi_c} \cdot T_c. \qquad 7.11$$

This is usually taken as representing the highest acceptable anode temperature. The efficiency of a reversible generator under these conditions is then seen to be the Carnot efficiency, $(T_c - T_a)/T_c$, which is equal to $(\varphi_c - \varphi_a)/\varphi_c$. In the example of the previous section, we find this efficiency to be 0·5 or 50%. As an illustration of how close to this we might expect to get, we can see from Figure 7.9 that at a solar intensity of 800W/m², the output power P_2 is about 230 W/m² for CR = 1 000, corresponding to an efficiency of nearly 29%.

In our example, the cathode temperature T_c was in the vicinity of 1500 K. The requirement 7.10 leads to a maximum anode temperature T_a of 750 K. If we could use the energy extracted from the anode to work a conventional heat engine working between say 700 K and a natural sink at about 300 K, we might, using the assumptions of the previous chapter, be able to recover about 28% of it as useful work. For the case cited above, the anode heat loss is 290 W/m², so that if 28% of this can be recovered, the total useful output becomes about 310 W/m². This represents an overall efficiency of nearly 39%. It remains to be decided in a practical case, whether this additional recovery warrants the very considerable complexity incurred in adding the secondary engine.

The overall efficiency of the combined thermionic generator and mechanical engine, in this example, is higher than that found for mechanical systems alone in the previous chapter. Yet both systems are subject to the limitations of the Second Law. The explanation is to be found in the maximum temperature in the two cases. Because, in a mechanical engine, the parts are subject to high stresses, we assumed that they could not be operated in practice at temperatures much in excess of 1 000 K. In the thermionic generator, however, a cathode temperature much higher than this is readily obtainable since the mechanical strength required of the cathode is not great. If we simply compare the Carnot efficiencies, assuming in each case a sink temperature of 300 K. we see that raising the source temperature from 1 000 to 1 500 K raises this limiting efficiency from 70% to 80%. This improvement is reflected in better efficiencies even in the less perfect systems with which we have to work in practice.

By now, it has become evident how stringent are the limitations set by thermodynamic restrictions on any system in which there is an initial heating stage. We might feel that further exploration of such systems would be unprofitable. However, low efficiency does not always mean that a system is of no value. If it can be made cheaply or operated simply, these features might be sufficiently advantageous to offset the low efficiency. It is for reasons such as these that attention is turning, for power generation in spacecraft and even for large-scale production, to another device, the THERMOELECTRIC GENERATOR. Although it must be said at this stage that its efficiency is lower even than that of other devices we have considered, it has the merit of simplicity and deserves attention as another potential converter for solar energy.

7.8 Thermoelectricity

In 1821, SEEBECK observed that if a circuit is constructed from two loops of dissimilar metals, a current passes round it if one of the junctions between the metals is kept hotter than the other. If the circuit is opened, a potential difference V develops between the two ends, which is proportional to the temperature difference. We then have

$$V = S(T_1 - T_2),$$ 7.12

where S is called the SEEBECK COEFFICIENT. Its value is a characteristic of the two materials used. The situation is illustrated in Figure 7.10a. This SEEBECK EFFECT can be considered to occur because electrons at the hot end have greater kinetic energy than those at the cold end, so that the diffusion of electrons towards the cold end occurs at a slightly greater rate than in the opposite direction. Excess electrons accumulate at the cold end until the repulsion of their Coulomb forces prevents a further increase. This charge separation produces the potential difference between the two ends of each part of the loop.

If the circuit is closed, electrons will move along both paths, but the different electron concentration in the two metals results in a nett current round the circuit. When a load is inserted in the circuit, as in Figure 7.10b, we have a generator. Energy passes into the hot junction and some is rejected out of the cold junction, the difference representing the useful work available in the load.

PELTIER noted in 1834 that if a cell is inserted into the circuit instead of the load, so driving a current round the system, energy enters at the low temperature junction and is rejected at the high temperature junction. This, too, is understandable in terms of the movement of electrons which have lower energy at the cold junction and higher at the hot junction. As might be

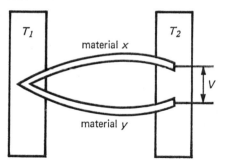

a)   Seebeck effect

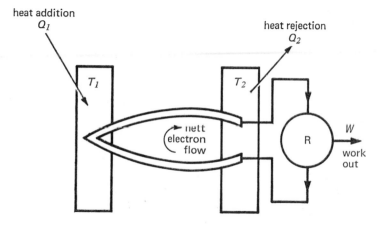

b)   Thermoelectric generator

Fig 7.10. Thermoelectric generation

expected, Peltier found that the energy transferred at a junction is proportional to the current flowing. At the hot junction,

$$Q_1 = \pi_1 \, I$$

and at the cold junction

$$Q_2 = \pi_2 \, I, \qquad\qquad 7.13$$

where the symbol π represents the PELTIER COEFFICIENT. This coefficient has a value which is found to depend both on the metals used and the temperature at the junction.

By reversing the direction of the current, the direction of the energy flows can be reversed. If the Peltier effects were the only ones occurring, the device would be completely reversible, so that the energy transfers would obey the thermodynamic relation

$$Q_2/Q_1 = T_2/T_1, \qquad 7.14$$

which we found for reversible heat engines in equation 6.10. Then, from equation 7.13, we would have $\pi_2/\pi_1 = T_2/T_1$ or $\pi_1/T_1 = \pi_2/T_2$, which means that π/T is a constant. In 1855, W. Thomson, later LORD KELVIN, first showed that this constant, which is a characteristic of the materials only, must be identical with the Seebeck coefficient, S. For, if the device is operating reversibly, that is, without losses, the current I must be very small. In that case, the potential difference across the cell or load would be near to V, given by equation 7.12. Thus, the power VI is equal to the difference ($Q_1 - Q_2$), giving, with equation 7.13,

$$SI (T_1 - T_2) = I(\pi_1 - \pi_2).$$

If $\pi_1/T_1 = \pi_2/T_2$, we then have Kelvin's result,

$$S = \pi/T. \qquad 7.15$$

The quantity S is readily measured and it will be convenient in our study to retain it as the only thermoelectric property. It is not a simple matter to predict its value from first principles, though we can quickly obtain an idea of its order of magnitude. It is very small in metals, about $10\mu V$ per K, which accounts for the general neglect of thermoelectricity until the discovery that it can be made very large in semiconductors. In these materials an electron has to be made to cross the energy gap before it enters the conduction band and has freedom to move and become a means of transferring charge. This is rather like the process of thermionic emission, discussed in Section 7.5, where an electron becomes a useful charge carrier only after surmounting the potential barrier representing the work function. It was stated in that section that those electrons which have surmounted this barrier, through thermal agitation, have an average energy in excess of the barrier height by about $2kT$. If we suppose similarly that in a semiconductor those electrons which are raised above the energy gap by thermal agitation have an excess energy of about $2kT$, then the difference in mean electron energy between the hot junction and the cold junctions would be about $2k(T_1 - T_2)$. If the nett current flowing is I, the power available is then $2kI(T_1 - T_2)$. If, by equation 7.13, we equate this to $(Q_1 - Q_2)$, we have

$$2kI(T_1 - T_2) = I(\pi_1 - \pi_2),$$

which would be satisfied if

$$\pi_1 = 2kT_1 \text{ and } \pi_2 = 2kT_2.$$

Then, using equation 7.15, we find that S has the value $2k$. This is about $1/6\,000$ V per K, or some $160\ \mu\text{V}/$ K, many times greater than the value for metals. In some materials, values up to $1\,000\ \mu\text{V}/\text{K}$ have been obtained. However, the successful exploitation of high Seebeck coefficients in semiconductor thermoelectric generators depends upon the unusual property of charge separation at the junctions between certain semiconductors. We shall defer consideration of this property until the following chapter, and meanwhile obtain some simple estimates of the output and efficiency of practical generators of this type.

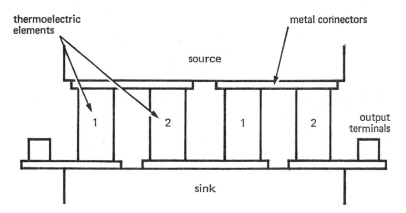

Fig. 7.11. Typical arrangement of thermoelectric generator

7.9 The thermoelectric generator

A typical arrangement of the conductors for a thermoelectric generator is shown in Figure 7.11. The elements are generally arranged in series because the output of practical devices is usually contrived to be at a potential difference of the order of 300–400 μV per unit temperature difference for each pair of elements. For a difference of 500 K, therefore, the output potential difference is only some 0·2V per stage.

In a practical device, certain irreversible effects occur. Heat transfer can take place from the source to the sink directly by conduction through the elements. As the current flows, Joule heating will occur within the elements. (Thomson showed also that when a current passes through a conductor in which there is a temperature gradient, the Joule heating is enhanced, because electrons moving from the hotter region have greater kinetic energy than appropriate to the cooler region and give this up to the oscillating ions of the 'fixed' lattice. We shall neglect this THOMSON EFFECT in our simplified study).

For any pair of thermoelectric elements, the rate of heat transfer by conduction will be proportional to the temperature difference between their ends, if we assume that no heat is lost from the sides. Then we could express this as

$$Q_h = K(T_1 - T_2) \qquad 7.16$$

where K depends upon the thermal conductivity of the materials and the areas and lengths of the elements (as seen later).

The Joule heating, upon passage of the current I, is given by

$$Q_j = I^2 R, \qquad 7.17$$

where R is the overall resistance of the elements, depending upon the resistivity of the material and the size and shape of the elements, as for thermal conductivity. If we assume again that no heat loss occurs at the sides of the elements, half the energy converted into Joule heat will pass into each of the junctions.

A power balance for the hot junction now yields

$$Q_1 = ST_1 I + K(T_1 - T_2) - \tfrac{1}{2} I^2 R \qquad 7.18$$

and that for the cold junction

$$Q_2 = ST_2 I + K(T_1 - T_2) + \tfrac{1}{2} I^2 R. \qquad 7.19$$

Then the useful power available in the load is

$$P_2 = Q_1 - Q_2 = S(T_1 - T_2)I - I^2 R \qquad 7.20$$

and the efficiency of energy conversion becomes

$$\frac{P_2}{Q_1} = \frac{S(T_1 - T_2)I - I^2R}{ST_1I + K(T_1 - T_2) - \frac{1}{2}I^2R}. \qquad 7.21$$

Now the value of this efficiency depends upon the current passing, the other quantities being constants for a given situation. The current drawn is at our disposal, for it depends upon the resistance inserted as a load in the external circuit. It is found that as this resistance is lowered from a high value, the efficiency at first increases and then decreases. The maximum efficiency is found to occur when the external resistance is

$$R_1 = R(1 + ZT_m)^{\frac{1}{2}}, \qquad 7.22$$

where $\qquad Z = S^2/KR \qquad 7.23$

and $\qquad T_m = (T_1 + T_2)/2$, the mean temperature. $\qquad 7.24$

If the external resistance is arranged to have the value given by equation 7.22, it is found that the efficiency, then a maximum, is given by

$$\frac{P_2}{Q_1} = \frac{n - 1}{n + r} \cdot (1 - r), \qquad 7.25$$

in which

$$n = (1 + ZT_m)^{\frac{1}{2}} \qquad 7.26$$

and $\qquad r = T_2/T_1. \qquad 7.27$

It will be seen that the maximum efficiency depends only upon the temperatures of the junctions and the quantity Z, defined by equation 7.23. Evidently, the larger Z is, the higher the efficiency, and we must enquire into the factors governing it.

If we suppose the cross-sectional areas of the elements in Figure 7.11 to be A and their lengths to be l, the conductance K in equation 7.16 will be $2kA/l$, k being the THERMAL CONDUCTIVITY of the materials, assumed about the same for both elements of a pair. Similarly, the total resistance, R, of the elements, if they are joined by good conductors at their ends, will be $2\rho l/A$, where ρ is the RESISTIVITY of the materials, again assumed similar for both elements. Then the determining quantity Z becomes

$$Z = S^2/KR = S^2/4k\rho \qquad 7.28$$

Now it can be seen that Z is a combination of properties of the materials used. Thus it is itself a property, and is called the FIGURE OF MERIT. Since it is defined here for the pair of materials forming a thermoelectric converter, it must be remembered that the Seebeck coefficient S used here is a property of the *pair*, not of the individual materials constituting it.

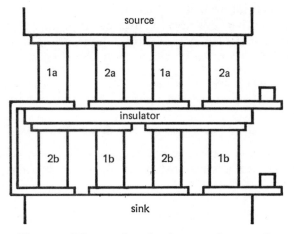

Fig. 7.12 Thermoelectric elements in cascade

Because its prediction from first principles is beyond our present scope, we may regard Z as a measurable property for any pair of thermoelectric materials and try to find those with the largest value. By a process known as DOPING, considered in the following chapter, it is possible to give certain semiconductors thermoelectric properties greatly superior to those of the metals. For the best known so far, Z has a value of the order of 0·003 per K at certain temperatures, but it falls off on both sides of the temperature giving such a value. Reasonably uniform properties can be obtained, however, by using several combinations of materials in a CASCADE, as illustrated in Figure 7.12. In a cascade, an intermediate temperature insulator is interposed between the thermoelectric elements having good properties at high temperatures and those having good properties at lower temperatures. The first might typically be made from a germanium-silicon alloy and the second from lead telluride. The presence of the intermediate insulator and the extra electrical contacts offsets to some extent the improved performance, and even with the best materials currently available, the effective figure of merit for such a combination is hardly better than about 0·000 5 per K.

If materials with this value of Z were used with a high melting-point storage system, perhaps operating at about 1 000 K, and with a sink temperature around 300 K, we would then have, from equations 7.24 and 7.26, a value of n of about 1.15. Then, from equation 7.25, the overall conversion efficiency would be only about 7%. As we saw in earlier sections, there are limits also to collection efficiency, even when using good quality mirrors with high concentration ratios, when the source temperature is around 1 000 K. Accordingly, we might conclude that we could not expect to recover much more than about 5% of the incident solar energy with a thermoelectric generator at the present time. At an incident flux of 800 W/m², this would represent an output of about 40 W/m², hardly better than that obtainable from a flat-plate collector and mechanical heat engine, described in Section 6.5. We can quite easily see why this is. Comparison of equations 7.25 and 7.27 with equation 6.9 shows that the $(1 - r)$ part of 7.25 is the Carnot efficiency for these temperatures. Then with a thermoelectric generator, we can only obtain a fraction $(n - 1)/(n + r)$ of the Carnot efficiency. For any practical case, this fraction is much less than the $\frac{1}{2}$ which we considered to be reasonable for a heat engine.

In spite of the poor efficiency at present obtainable, interest in thermoelectric generators continues at a high level. Today's efficiencies are ten times better than those of a few decades ago, and it might be expected that the continuing search for better materials will yield further improvements. For example, by equations 7.24, 7.25 and 7.26, we can see that if an effective figure of merit of 0·005 per K could be obtained over the temperature range 300–1000K, the efficiency would not be a mere 7%, but a competitive 31%.

It is worth noting, however, that the variation of figure of merit with temperature can bring about a situation which is favourable to the use of a flat-plate collector with a thermoelectric generator, as shown in Figure 7.13. Although the maximum temperature will now be much lower, we may find a pair of thermoelectric materials with a high figure of merit for

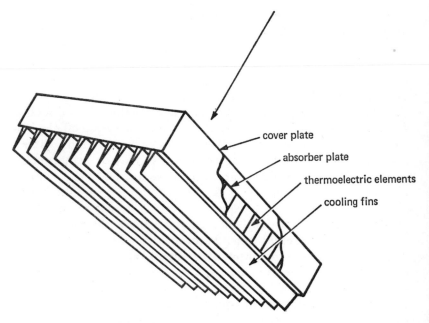

cover plate

absorber plate

thermoelectric elements

cooling fins

Fig. 7.13 Thermoelectric generator with flat-plate collector

the particular temperature range used, which is now much narrower. For example, with T_1 only 400 K, but using materials with $Z = 0.002$ per K, we obtain a value of n of about 1·27, giving an overall efficiency of nearly $3\frac{1}{2}\%$. Since we could get such a working temperature without the complication of a concentrator and the need for a tracking device to follow the sun, this seems quite promising. However, this efficiency is based on the utilisation of the energy actually entering the generator. We saw in Section 5.3 how the collection efficiency of a flat plate collector falls as power is taken from it. It is easy to work out the performance of a combined collector-generator system, just as we did for a combined collector-engine system in Section 6.5. If we do this, we find that, assuming a value of the figure of merit Z to be 0·002 per K, the output when working at maximum efficiency is about as shown in Figure 7.14. At an incident flux of 800 W/m², the output, using an absorbing surface with neutral characteristics, is only about 6 W/m². Thus the overall efficiency is about 0·75%. (Such a device, using an alloy of bismuth and antimony as one element and zinc antimonide as the other, was built as long ago as 1954 by Dr. Telkes, to whom we referred in Section 5.6. With this, she obtained an overall efficiency of about 0·6%). With the selective absorber, the expected performance is somewhat better, with an overall efficiency at the same operating point of about 1·4%.

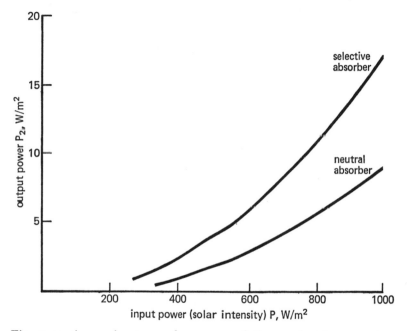

Fig. 7.14 Approximate performance of thermoelectric generators with flat-plate collectors

What we have seen above, and in the previous chapters, shows that whatever we do, thermodynamic limitations of a fundamental kind will prevent us from recovering more than a minor part of the solar energy via devices in which the first stage is the absorption of this energy, leading to an increase in temperature. Even making liberal assumptions, we find that we cannot hope to obtain more than about 40% of the energy in the form of work, and know that further practical difficulties will reduce the proportion to much less than this in most cases. Thus, we have reached a stage at which we would expect it to be unprofitable to examine other devices for energy conversion which also involve an initial heating stage. (Another of these now being seriously studied for power generation on a large scale, for instance, is based on an effect known as MAGNETO-HYDRODYNAMICS or MHD). In the next two chapters, therefore, we shall turn to other possibilities. Their essential feature is that they seek to obtain the energy of the incoming radiation photons *without* any appreciable temperature rise of the bulk of the receiving surfaces.

8

PHOTOELECTRICITY

The art of discovery advances with invention.

FRANCIS BACON (1561–1626)

IN Chapter 7, we saw that one of the events that helped to gain general acceptance for quantum theory was the explanation of the PHOTOELECTRIC EFFECT by EINSTEIN (1905). In this, first reported by HERTZ in 1887, the interaction of a light photon with an electron gives the latter enough energy to escape from the parent metal. If enough electrons could be made to do this, they might be collected on another metal surface and caused to pass through an external circuit on their way back to the emitting surface, just as in the thermionic generator previously described. There would however, now be no initial heating stage; the energy is given directly to the electron. We would expect the efficiency to be high, not being subject to thermodynamic limitations.

In this chapter, we first examine this process, as realised in the PHOTOEMISSIVE GENERATOR, and then move on to consider the more practicable case of photoelectric generation in the SEMICONDUCTOR DIODE. We shall see how, by quite recent developments, the long-standing dream of circumventing the restrictions of thermodynamics has now been realised.

8.1 *The photoemissive generator*

Figure 8.1 shows the electrodes of a hypothetical photoemissive generator. Radiation photons reach the cathode by passing through the anode, which is shown here as a grid of wires. To avoid a serious space-charge effect (described for the thermionic generator in section 7.6) the electrodes are placed very close together.

When a photon arrives at the cathode, it may be absorbed by an electron, which is thereby raised to a higher energy state. If the energy increase exceeds the work function for the cathode

183

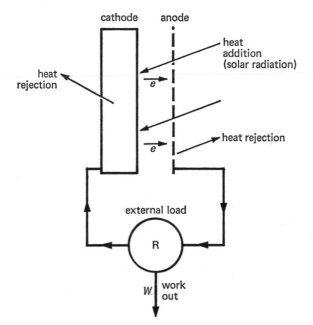

Fig. 8.1 Photoemissive generator

material, the electron might escape from the surface; this is what we require. Not all photons have enough energy to produce this effect. We saw in Section 4.1. that the energy of photons is related to the frequency by the expression

$$E = h\nu = h\upsilon/\lambda, \qquad 8.1$$

first given as equations 4.1 and 4.2. For solar radiation, with wavelengths around one micrometre, this expression reduces for convenience to the form

$$E = 1.24/\lambda, \qquad 8.2$$

when the unit of energy is the eV and that of the wavelength is the μm. Only photons with energy greater than the work function can cause electrons to leave the cathode. For $\varphi_c = 2$ eV, as for our example of a thermionic generator, this would mean, by equation 8.2, that only radiation with wavelength *less* than 0·62 μm would be capable of doing this. In a typical spectrum of solar energy near the ground, only about 30% of the radiation has wavelengths less than this value. Thus, on this count alone, no photoemissive device could have an efficiency greater than 30% if the cathode work function were 2 eV. Further, we saw in the case of the thermionic generator that the energy drop of electrons on arrival at the anode is $\varphi_c - \varphi_a$, and that this is given up to the oscillating ions of the anode and radiated away. Then, of the maximum possible 30%, only a fraction $(\varphi_c - \varphi_a)/\varphi_c$ could be available for useful work. With $\varphi_a = 1$ eV, as in our example, this would be $\frac{1}{2}$ and the maximum efficiency has fallen to 15%.

Other choices of electrode materials, with different work functions, might yield a more attractive result, but there remains an even greater restriction. When we were dealing with the thermionic generator, we saw that significant emission of electrons from the cathode would only take place when the temperature was high enough for many electrons to have energies exceeding the Fermi level by an amount φ_c. When this happens, there are many more electrons with energies only slightly less than this. But in a photoemissive generator, only those electrons directly excited by radiation have useful energies; the remainder, and the ions of the crystal lattice, have energies appropriate to the cathode temperature only. An excited electron soon loses its energy by interactions with these lower-energy neighbours, so that only a very few, excited near the cathode surface and scattered in an outward direction, actually manage to escape. It is for this reason that the efficiency found in practice for this kind of device is not even 15%, but nearer to 0·15%! We have to conclude that it would be better to use a concentrator for the radiation and let the cathode heat up until we have a thermionic generator again.

This unpromising outlook has no doubt been responsible for the neglect of photoemissive devices in studies of solar power generation, though it is probable that, by ingenious design of the electrodes and use of cathodes in the form of thin films, the efficiency could be raised substantially. However, devices of this kind have been obscured by the relatively spectacular success of the PHOTOELECTRIC or PHOTOVOLTAIC GENE-RATOR, which we shall consider next.

8.2 *The photoelectric generator*
In Figure 7.2. were illustrated the electron energy band struc-

tures of representative materials. It will be recalled from the discussion accompanying the figure that electrons can only be caused to move, and thus to constitute a current, if they are able to acquire some energy. This means that they must have a permissible state to pass into. This can happen easily if they are in a partly-filled or conduction band, where there are plenty of available states. Electrons with states in the filled or valence band, however, cannot move unless they are given enough energy to cross the forbidden band and enter a state with a level in the conduction band. In the intrinsic semiconductors, such as germanium or silicon, the energy gap to be crossed is of the order of 1 eV. This is about the same as the energy of photons of light. The functioning of the photoelectric generator depends upon the raising of electrons in semiconducting materials into conducting states by the energy obtained from the absorption of photons of sunlight. Each electron so raised will have an energy relative to its original state equal to the energy gap. However, in the ordinary way, the electron would only remain in this high-energy state for a short time before re-combining with an ion and relinquishing the energy to lattice vibrations or re-radiation. Where energy has passed into the ion lattice of the solid, whose vibrations are thereby invigorated, we say that the temperature has risen. This is just what we want to avoid. The successful operation of a photoelectric generator depends upon causing the excited electrons to pass out of the semiconducting material and to give up their excess energy usefully in an external circuit before being returned to the ground state. We shall consider later how this can be done, but first it will be instructive to see just how profitable such a system could be. By preventing the conversion of the solar energy first into heat, we may avoid thermodynamic limitations, but by now we shall not be surprised to find that Nature has placed other restrictions in their place.

Equation 8.2. showed how the energy of radiation photons diminishes with increasing wavelength. Evidently, there comes a value of wavelength beyond which the energy of a photon is too low to cause an electron to cross the energy gap. For silicon, for example the energy gap is about 1·1 eV at ordinary temperatures, which, from equation 8.2, is equal to the energy of a photon of wavelength about 1·1 μm. At sea level, roughly 20%

of the solar radiation has wavelengths longer than this and therefore cannot be utilised in a device employing silicon. Unfortunately, the radiation with shorter wavelengths cannot be fully used either. Because electrons can only move to permitted states, only an amount of energy close to the energy gap can be retained for any length of time. If the incoming photon has more energy than this, the excess is quite quickly transformed by interactions into lattice vibrations, that is, into internal energy. Thus, as we saw earlier, a photon with wavelength of about 0·6 μm has about 2 eV of energy, but electrons in silicon can still only take up 1·1 eV of this, the remainder being not only wasted, but becoming a nuisance, because it merely raises the temperature.

We can now make a rough estimate of the greatest proportion of solar energy which could be converted into electrical energy by a perfect photoelectric generator. It depends somewhat on the distribution of energy with wavelength in the solar spectrum, which varies with latitude and meteorological conditions, as shown in Figure 2.9 and discussed in section 2.5. We may take as representative the data of Table 8.1, roughly the distribution found in clear conditions near the tropics.

TABLE 8.1

Maximum possible energy conversion in silicon (cut-off at 1·1 μm)

Wavelength interval, μm	Proportion of solar energy in interval, %	Fraction converted in interval	Proportion recovered, %
less than 0·3	0	—	—
0·3–0·5	17	0·36	6
0·5–0·7	28	0·55	15
0·7–0·9	20	0·73	15
0·9–1·1	13	0·91	12
greater than 1·1	22	0	0
	100		48

In the third column of the table, the fraction converted is simply taken to be the ratio of the wavelength at the centre of the interval to the cut-off wavelength corresponding to the energy gap. (It will be seen from equation 8.2, that if the photon energy at wavelength λ is $1\cdot24/\lambda$ and the energy gap is such that a photon of wavelength λ_g just has that energy, the useful fraction of the energy at wavelength λ is $1\cdot24/\lambda \div 1\cdot24/\lambda_g$, which is λ/λ_g. All photons with wavelengths greater than the CUT-OFF WAVELENGTH, λ, cannot energise an electron at all.)

By repeating this kind of exercise with different values of the cut-off wavelength, we find that silicon, with λ_g about $1\cdot1\,\mu m$ is about the best material, although the greatest possible efficiency is about 45% for a range of values of λ_g on either side of this value. We conclude that no device employing electron excitation in this way could have an efficiency greater than about 45%. Disappointing as this is, we find in practice that the efficiency of actual devices is even less. To understand why, we must look in more detail at the methods used to recover the energy acquired by the electrons.

8.3 Doped semi-conductors
Separation of the excited electrons from the ions which they have left is usually accomplished by using the properties of semiconductors which have been DOPED. Doping consists of introducing into the lattice structure of an intrinsic semiconductor some impurity atoms of a different element, having more or fewer valence electrons than those of the bulk material. These impurity atoms can be introduced, for example, by diffusion during contact at elevated temperatures. The effects are best illustrated by a specific example.

The semiconductor silicon has four valence electrons in its outermost shell of orbits. In a crystal, these electrons are shared with those of four neighbouring atoms, joining them together by what are called CO-VALENT BONDS. Each atom of every pair of atoms involved provides one of the two electrons forming the bond. As a result, the regular lattice of the crystal is built up in a particular geometric pattern (which can ultimately be seen in the shape of large crystals). We cannot properly illustrate such a three-dimensional arrangement, and cannot draw the electrons at all, but the method of bonding is shown for conceptical purposes in the sketch of Figure 8.2.

Now it is found that small quantities of IMPURITY ELE- MENTS, having three or five valence electrons, can be

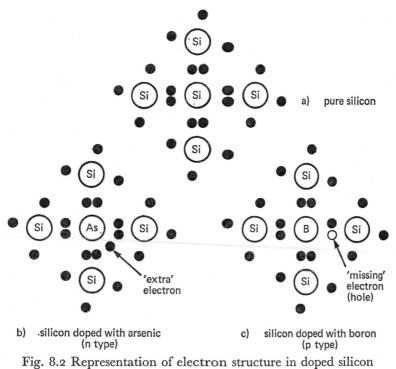

a) pure silicon

'extra' electron

'missing' electron (hole)

b) silicon doped with arsenic
(n type)

c) silicon doped with boron
(p type)

Fig. 8.2 Representation of electron structure in doped silicon

introduced into the silicon crystal without seriously distorting the lattice. If they are not much different in size, they can be tolerated in concentrations up to one part in a million or so. If atoms of, say, arsenic are introduced, they take up positions on the regular lattice like silicon atoms, but they have an extra electron. This electron is not required in the valence bonds and can be regarded as a free carrier of negative charge. The crystal as a whole is still electrically neutral, because there is an extra positive charge in the nucleus of the impurity atom to balance each free electron, but this positive charge is bound in position on the lattice. A material doped in this way is called an N-TYPE SEMICONDUCTOR, to indicate that the free charge carriers in it are electrically negative.

If we introduce instead impurity atoms of, say, boron, or aluminium, which have only three valence electrons, they are also taken into the crystal lattice. But now there are insufficient electrons to form the required bonds. The vacant space where an electron ought to be is called a HOLE. Now it happens that an electron near a hole of this kind can move into it and satisfy the empty bond. This requires no energy change for an electron which was previously bound. In moving, it vacates a bond nearby, leaving a hole there. The effect is as if the electron and the hole had exchanged positions. In the crystal as a whole, valence bonds filled by electrons are the normal thing and only the holes are noticeable. When an electron moves into a hole, the hole appears to have moved in the opposite direction and the material is behaving as if it had freely-moving holes. The electrons moving into the spaces previously 'occupied' by the holes move, in general, only one space at a time and then remain fixed, whereas the holes are able to move continually, exchanging places with a series of electrons. Relative to the rest of the crystal, these holes, representing an absence of the negative charge of the electron, are like objects with a positive charge. It is thus convenient to think of them as free POSITIVE CHARGE CARRIERS. A material doped in this way is called a P-TYPE SEMICONDUCTOR. Again, the crystal as a whole is neutral because there are fewer positive charges located in the nuclei of the impurity atoms than there would have been in those they have replaced.

Although the proportion of impurity atoms in doped semiconductors is very small, the number of free charge carriers provided by them is much greater at ordinary temperatures than those freed by incident radiation or by thermal fluctuations. Carriers with charge of the sign produced by the impurity are therefore known as MAJORITY CARRIERS, and those of either sign liberated by radiation or increased temperature, MINORITY CARRIERS.

We have now arrived at a simple picture of doped semiconductors of two basic types: n-type, having fixed positive and free negative charges, and p-type, having fixed negative and free positive charges. At the same time, there will be in both types minority carriers released by radiation or thermal effects. Hitherto, we have thought only of the electrons in this category. We should note that each electron, when energised enough to escape from an ion and cross the energy gap, must leave a hole behind it. From the previous discussion, we can see that in this case, both the electron and the hole are free to move.

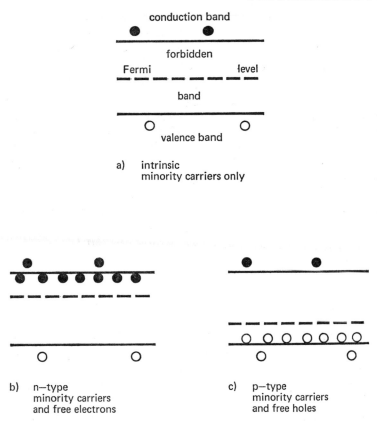

a) intrinsic
    minority carriers only

b) n—type
    minority carriers
    and free electrons

c) p—type
    minority carriers
    and free holes

Fig. 8.3 Simplified representation of energy levels in semiconductors

Before considering how these features can be exploited in an energy converter, we shall need to examine the effects of doping on the representative or Fermi level, introduced in Section 7.3. We saw in Figure 7.4 that this level is in such a position that, on an increase in temperature, the increase in number of electrons having energies greater than the Fermi energy is equal to the decrease in number of those with energies below it. Now in an intrinsic (un-doped) semiconductor at low temperatures, all the electrons are in the valence band, completely filling its energy levels. An increase in temperature causes a few electrons to pass the forbidden band and take up states in the conduction band, leaving an equal number of holes in the valence band. It follows that for such a material, the Fermi level is located near the centre of the forbidden band, as shown in Figure 8.3 (a). If this semi-conductor had been doped so as to become n-type, the excess electrons, being free to move, must have energies close to those of the conduction band. The Fermi level is thus higher in this material, becoming closer to the bottom of the conduction band, as shown in Figure 8.3 (b). Similarly, the introduction of p-type doping produces a material in which the Fermi level is nearer to the top of the valence band, as shown in Figure 8.3 (c). These diagrams are intended to show the electronic energy-levels, so that the representation of holes on them is improper. However, they have been included to make clearer the disposition and relative freedom of both types of charge carrier as an aid to understanding this rather complicated situation.

8.4 The junction diode

If a piece of n-type and a piece of p-type material are placed in contact, the majority carriers, comprising free electrons in the former and free holes in the latter, can drift across the junction into material of the opposite kind. There they can meet and neutralise each other and, in effect, vanish. We would then find in the vicinity of the junction, a region in which only the bound positive and negative charges remain, as shown in Figure 8.4. Thus, there has developed on either side of the junction a narrow zone, known as the DEPLETION LAYER, in which, instead of the neutral condition existing elsewhere, there is excess positive charge in the n-type material and excess negative charge in the p-type material. This will oppose the drift of charge carriers and, in the absence of other effects, would build up until no more majority carriers could cross. However, the difference in potential represented by the excess charge is such as to *assist* the minority carriers across. Electrons in the p-type material and holes in the n-type material are now encouraged to cross, whilst the holes and electrons can cross only with difficulty, whenever thermal fluctuations happen to give them enough energy. There is thus a two-way motion of both holes and electrons across the junction. Left to itself, the device will attain a condition of equilibrium in which equal

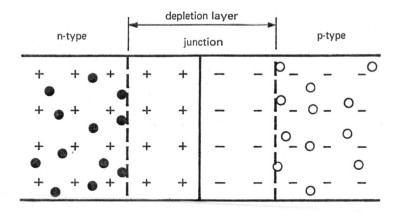

± represent fixed charges

Fig. 8.4 Majority carriers in vicinity of p–n junction

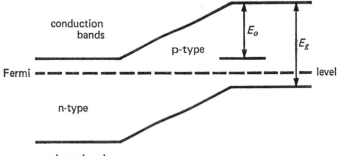

Fig. 8.5 Energy levels at p-n junction on open circuit
(no illumination)

numbers of carriers are passing in each direction, a small proportion of the majority carriers with difficulty and the minority carriers with ease. In this condition, there is no nett flow of electrons across the junction, so that there is no real current. This means that the accommodation that takes place between the two materials when they are placed in contact is such as to bring the Fermi levels on both sides to the same value. The corresponding relative positions of the electronic energy bands in the materials near the junction are thus as shown in Figure 8.5. The height of the potential 'hill' along each band edge is E_0, as shown. From Fig. 7.11 it can be seen that E_0 is the difference between the distances of a given band from the Fermi level in the two types of material. Evidently E_0 is, in this case, less than the energy gap E_g.

We now have a device known as a junction DIODE ('two electrodes'), whose most interesting characteristic from our present point of view is that it has the capacity of assisting electrons to cross from the p side and holes to cross from the n side, whilst restraining the carriers of the opposite kind. If we now allow radiation to fall on the material near the junction, it will produce electron-hole pairs there. The action of the junction will then be to separate these, which is just what we need in an energy converter. We then have to lead the electrons from the n side, take them through an external circuit and return them to the p side, where they can meet and recombine with the holes passing the other way. (The concept of the hole only has meaning in the semiconductors, so that we do not think of holes passing round the external circuit.)

The passage of current through the external circuit requires a suitable potential difference and this must be related to the energy levels at the junction. This effect is considered in the next section.

8.5 *Output and efficiency of photoelectric generators*
If we have an external circuit, there will be points, the CONTACTS, at which it is attached to the semiconductors. The external circuit will usually be made of metal and at the metal-semiconductor contacts, new depletion layers will develop by electron diffusion (in the semiconductors only) in a manner similar to the formation of those at the junction. These 'half-

junctions' have the effect of ensuring that no potential differ-ence appears across the contacts in the equilibrium state we have described. (Whatever the nature of the external circuit, in fact, the behaviour at each of the terminals will be roughly equivalent to half that of a direct second junction of the two semiconductors). If we do not, for the moment, switch on the external circuit, the effect of the radiation being allowed to fall on the generator is to disturb the equilibrium state. As we have seen, electrons will pass into the n side and holes into the p side, so that the n side charges up relative to the p side. This will have the result of reducing the height of the potential hill and a new equilibrium position will be reached, accompanied by the appearance of a potential difference across the contacts. This is called the OPEN-CIRCUIT VOLTAGE. For low values of illumination, this voltage is small because the number of electron-hole pairs produced is much less than the number of majority carriers present due to doping. As the illumination increases, however, the open-circuit voltage V_0 rises until a point is reached at which it approaches E_0, when, because of the predominance of these new carriers, the potential hill at the junction is all but eliminated.

The other extreme of operating conditions is the SHORT-CIRCUIT case. If the two semiconductors are connected by an external wire having negligible resistance, the Fermi levels are equal at the ends and must remain so throughout the device. Photo-generated electron-hole pairs are now separated vigor-ously at the junction, which has the full potential hill across it. The short-circuit current I_s, which is passing in this condition, is directly proportional to the intensity of the incident radia-tion. However, since there is no resistance in the external circuit, no work is being done there.

Evidently, we need an intermediate case, in which the potential drop across a load in the external circuit will be greater than the zero value of the short circuit case, and the current will be greater than the zero value of the open-circuit case. If the external circuit has any resistance, the potential difference V required to cause a current to pass must be obtained by a difference in the Fermi levels on the two sides of the junction. These levels and the relative position of the band edges are then as shown in Figure 8.6. The height of the potential hill across the

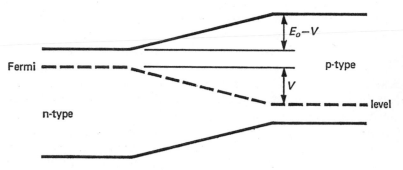

Fig. 8.6 Energy levels at p-n junction with external resistance
(junction illuminated)

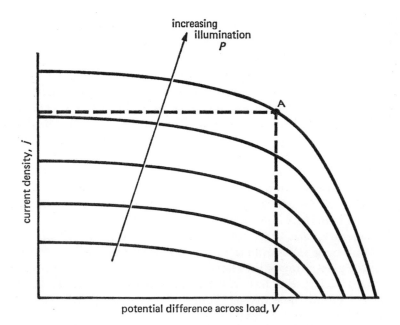

Fig. 8.7 Typical characteristic curves for photoelectric generator.

junction has now been reduced from the value E_0 in the short-circuit condition to $(E_0 - V)$. Thus the junction cannot now be so effective in separating the radiation-generated pairs as in the short-circuit case. We then find that the current density j for a given radiation intensity P falls with increasing V. This is because some of the carriers which were formerly restrained can now manage to get 'uphill', crossing the junction the 'wrong' way. The general form of the relation between P, j and V is illustrated in Figure 8.7. A theoretical analysis of the very complex behaviour of the illuminated junction, leading to curves such as this, is beyond our present scope. (Indeed, it might be said that it has not yet been entirely satisfactorily achieved from first principles anywhere). But enough has been said to explain the main features of the behaviour of photo-electric generators, and will allow us to make a rough estimate of their actual efficiency.

Figure 8.7 showed diagrammatically the relation between current and potential drop across the load. We are interested in the *power*, which is the product of these quantities; jV is the power produced in the load per unit area of surface exposed to radiation. It will be seen that the operating condition for which this product is greatest is when a rectangle inscribed within the characteristic curve has the biggest area, since the area of the rectangle is $j \times V$. Such a case has been marked in Figure 8.7, with the operating point at A. In this condition, the potential drop across the load is rather less than E_0, and from the previous discussion we might expect it to be about $\frac{1}{2} E_g$. Because of the slope of the curve, the current density j will not be much less than the short-circuit value j_s. This is the highest current produced by photon-generated electron-hole pairs. Its value depends on how effectively the pairs are separated without recombining, a quantity which cannot be calculated very accurately at the present time. Evidence from actual devices, however, shows that by careful design, it is possible to recover more than 80% of the pairs generated in some materials. At the operating point, we might be getting about 70% of them. The power available externally would then be about 0·5 × 0·7, or about 35%, of the ideal value calculated in Section 8.2, where we assumed maximum conversion efficiency of those photons which are useable. There we found that with a typical solar energy spectrum, about 45% of the incident radiant energy could be converted. After our more detailed examination, we would now expect about 35% of this, or about 16% of the incident energy. Many generators have, in fact, been produced with efficiencies of about 15%, so our estimate is realistic.

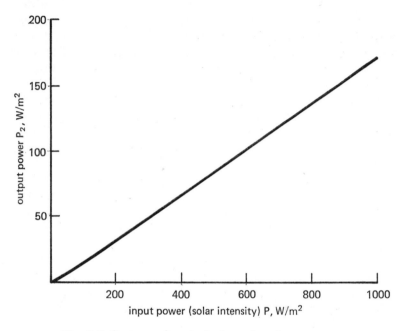

Fig. 8.8 Output of typical photoelectric generator

The output of a typical photoelectric generator exposed to tropical sunlight and well matched to its load, would thus be expected to be as shown in Figure 8.8. There is some fall in efficiency at low illuminations because, as shown in Figure 8.7, we cannot there obtain quite so high a potential drop across the load. Again we have a low-voltage device; for silicon generators, V will be about 0·6 volts only. At a typical input of 800 W/m², we would expect in practice to obtain about 130 W/m² of useful power. The fate of the energy which is not employed usefully is important. Transformation into lattice vibrations must be avoided, since these encourage carriers to surmount the potential 'hill' and pass the wrong way across the junction. Since lattice vibrations are a symptom of an increased temperature, this is equivalent to saying that we must keep the temperature down. The effects of not doing so are severe. In a typical case, an increase in operating temperature from 20 to 100°C would reduce the output by more than a third. Radiation in parts of the spectrum which cannot be used efficiently can be reflected away by suitable coatings, but there is inevitably some conversion into heat within the generator and arrangements must be made to conduct or radiate this away as effectively as possible.

Photoelectric generators built so far (the 'solar cells' of spacecraft and other devices carrying their own power supply) have used semiconductors cut from single crystals. A typical arrangement is shown in Figure 8.9. We end this chapter by considering one or two feasible methods of improving the performance of these generators.

8.6 Development of photoelectric generators

We have seen that, in spite of our optimism over quantum-conversion devices, the efficiency of simple photoelectric generators does not at present exceed that of other competitive systems, employing mechanical heat engines and thermionic converters. It appears from the discussion that the low efficiency is due mainly to two effects: the high fraction of the energy of some of the incident photons which cannot be communicated to electrons and the low fraction of E_g represented by the potential difference V, which can be developed across an external load. It is likely that novel designs now being actively pursued will reduce the losses caused by both these effects. The second effect is least for highly-doped semiconductors with a high energy gap. For these the number of carriers crossing the junction the 'wrong' way is much reduced and the potential 'hill' is retained at higher values of the fraction V/E_g. Maximum power can then be obtained at V/E_g values appreciably greater than the 0·5 we assumed—perhaps up to 0·7 or more. From the method of section 8·2, however, we can see that when E_g is high, a smaller portion of the solar energy can be communicated to electrons than when it is around the optimum value of about 1·0—1·4 eV. Studies are in progress aimed at offsetting this by using compound devices of the kind shown in Figure 8.9. The idea is similar to the cascading of thermoelectric generators, shown in Figure 7.12.

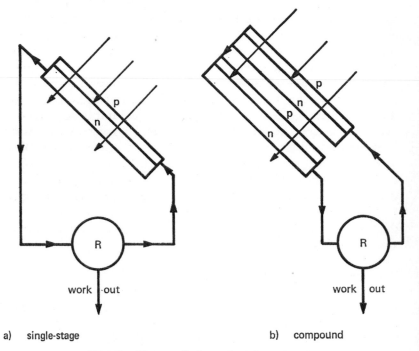

a)   single-stage   b)   compound

Fig. 8.9 Types of photoelectric generator

The radiation first falls on a high energy gap cell, having high efficiency for that part it can capture. Photons with energy lower than E_g cannot be used and in fact, the material is transparent to them. These photons pass through the first stage and into a second, with lower energy gap. Here the capture properties are good, though the overall efficiency is lower than in the first stage. The combination has better total efficiency than any single-stage device. For instance, suppose that the first stage has a cut-off wavelength of 0·7 μm, and the second stage one of 1·1 μm, as before. Using the data of Table 7·1 for a rough calculation, we find that the proportion of the total solar energy recovered in ideal circumstances is about 34% in the first stage and about 27% in the second. Now if the first stage can be operated with $V/E_g = 0\cdot7$ and at 80% of short-circuit current, its working output becomes 0·7 × 0·8 × 34 = 19%. For the second stage, we recover 0·5 × 0·7 × 27 = 9%, using the figures we employed previously. Thus the total recovery is 28%, nearly twice the efficiency of the single stage device. A further probable development is an integrated generator so doped that it has an energy gap that varies from a high value near the exposed surface to a low value deeper inside the material. This and other developments hold a definite promise of improved overall efficiency and it is widely believed that photo-electric devices will ultimately be capable of efficiencies greater than for any other system, perhaps in the range 50—60%. Efforts are continuing also to develop devices much cheaper than those produced so far, which are extremely expensive.

The cost of silicon photocells at present is a few dollars per square centimetre of exposed surface. This high cost is understandable when consideration is given to the extraordinary care required in the preparation of the semiconductor elements. We saw that doping involves the deliberate introduction of impurity atoms at concentrations up to about one part per million. Before this can be done, the basic material must be cleared of other impurities down to a level of about one part in a hundred million. Hardly any other use calls for material in such a high state of purity. Moreover, for the highest photoelectric efficiency, it is necessary that the semiconductor shall have no irregularities in its lattice structure, which in practice means that it has to be cut from a single crystal.

Silicon is a hard, yellow non-metal, obtained from silicates which occur in great abundance in igneous rocks, or from silica (SiO_2) which is the main constituent of some sands. The raw material is, therefore, very common; in fact silicon is the second most abundant element, after oxygen. To obtain the highest purity, however, many stages of refinement are necessary. The final reduction of impurities is usually obtained by a method known as ZONE-REFINING. In this, a short molten zone is formed near one end of the ingots by a surrounding coil through which a high-frequency current is passed. This zone is then made to travel very slowly along the ingot. It is found that this causes impurities to be swept along with the molten zone, leaving a substantial part of the ingot in a very pure state. The ingot is then broken up and the purest part remelted with the required amount of the doping element. Photocells are obtained from single crystals of this doped material.

Single crystals up to a few centimetres wide can be obtained by very gradual solidification of material on a small 'seed' crystal mounted on a probe dipping into a crucible of the molten element. The probe is very slowly withdrawn, bearing with it a growing column of material which has the perfectly regular structure of a single crystal. Very thin slices of this crystal are cut by sawing to obtain the wafers needed for photocells. A common method of producing the p–n junctions is then to diffuse atoms of the seond doping element into the surface of the cells by holding them at high temperature in an atmosphere of these atoms. The diffused layer has to be very thin (a few μm), otherwise the incident light will be unable to penetrate to its lower edge, which forms the junction. Finally, the cell is plated to form areas to which the electrical contacts can be soldered to obtain the lowest possible resistance.

It is worth noting, in passing, that the cost of semi-conducting materials for thermoelectric generators is much lower than for photocells. This is partly because the levels of doping needed to give good thermoelectric properties is as high as one part per thousand. Correspondingly, the impurities in the undoped material need only be reduced to about one part per million. Moreover, thermoelectric elements need not be made from single crystals. They can be cast, or fabricated by SINTERING. In this process the material is ground up into particles, which, when pressed together at high temperatures (but below the melting point) become fused into a homogeneous block.

It is recognised that the high energy-conversion efficiency of photoelectric cells is gravely offset by the high cost of manufacture. One way of making them effectively less expensive is to use a focussing solar collector with only a small cell at the focus. For all practical purposes, the output of the cell remains proportional to the intensity of illumination, without signs of saturation, up to concentration ratios of at least 10. As the output current rises, so do the losses due to Joule heating. As we saw in equation 7.2., these losses are proportional to the resistance and the square of the current. Special contacts in the form of grids of wire have to be used to keep the resistance of the cell to a minimum, but the Joule losses and the general problem of keeping the cells cool severely limit the useable CR in practice.

An alternative is to use doped silicon or other semiconducting material in a polycrystalline form. This avoids the slow and expensive stage of growing single crystals. The crystal size in an ingot can be controlled within limits by controlling the rate of cooling so that a slice used for a photocell can contain a small number of boundaries between adjacent crystals. The distortion of the lattice structure near these boundaries has the effect of lowering the energy gap and hence the efficiency of the cell. The reduction of efficiency to about 4% for polycrystalline silicon is accompanied by a reduction in cost of cells by a factor of two or three, which leads to about the same cost of power generated. However, many other substances have been shown to produce the photovoltaic effect since the first report of this by ADAMS AND DAY in 1877. A layer of copper oxide on copper is one of the best known, but much interest is at present centred on cadmium sulphide, which is much used in photoelectric expo-sure meters for photography. This and other promising sub-stances can be prepared in single crystal form, but show no advantage over silicon, which as we have seen, has an energy gap close to the optimum value. When CdS is laid down as a thin film, however, the average energy gap is lowered and it becomes close to the optimum value. Thin films can readily be obtained by VACUUM DEPOSITION. In this, the material is heated in an evacuated vessel so that molecules evaporate and then condense onto the required surface. The photovoltaic efficiency for CdS deposited in this way on copper is said to be about $3\frac{1}{2}\%$. Thin-film photoelements have not yet received as much attention as single-crystal types, largely because most of the effort in this field has been directed towards the requirements of space vehicles, where the highest efficiency is demanded, without regard to cost, so that weight is kept to a minimum. In the realm of solar energy conversion, the relative value of thin-film photo-cells might be enormously greater. Though there is not much experience with them yet, it has been said that in large-scale production, the cost of thin-film CdS cells might be measured in dollars per square metre rather than per square centimetre. If this were really so, the rather unfavourable present position of photocells for power generation would be quite reversed.

The high efficiency expected of the photoelectric generator is due to the direct transmission of the energy of radiation photons to the electrons of the device. The energy must then be used immediately or stored in some other device, such as a battery, which is an electrochemical cell. In the next chapter, we shall look briefly at the possibility of storing the energy in chemical or biological systems directly without the employment of a separate photoelectric generator. Since quantum changes are again involved, we might expect such processes to be highly efficient.

9

PHOTOCHEMISTRY AND PHOTOBIOLOGY

The force that through the green fuse drives the flower . . .

DYLAN THOMAS (1914–1953)

IT is likely that nearly everyone who has ever lived has experienced the reactions produced in the skin by sunlight, leading to suntan and sunburn. Many other light-induced chemical changes have been noted from time to time. The fact that the bleaching of dyestuffs was caused by light has been known and exploited for centuries and as early as the 18th century, the blackening of certain salts of silver on exposure to light had been noted. A further effect was revealed in 1839 when A. C. BECQUEREL observed that the potential difference across a chemical cell, with which he was experimenting, changed when light was allowed to fall on one of the electrodes. From these beginnings has grown the field of study which became known as PHOTOCHEMISTRY and, more recently, RADIATION CHEMISTRY.

Though some of the results of this study have been of great importance to mankind—such as the development of photography—large-scale practical applications have so far been rather few in number. No doubt this has been partly due to the experimental difficulties inherent in the subject; only recently has it been made possible to study the early stages of photochemical reactions, which last perhaps for as little as a millionth of a second. Now that these early stages are beginning to be understood, the subject is developing extremely rapidly. It will not be appropriate for us to examine these developments closely here. All that is necessary is to look at some of the principles of the subject and to show why major contributions to the field of energy conversion might be expected to come from it in the near future.

Sunburn is not the most important instance of a photobiological reaction. Nature has always held before us the supreme example of PHOTOSYNTHESIS, in which green plants and certain other organisms construct complex organic molecules from water and carbon dioxide, using the energy from the sun. Without this source of food, animal life would have been impossible. Now that the continuance of this life is faced with unprecedented difficulties, it is natural for us to consider the possibility of controlling the photosynthetical process by new kinds of husbandry. We shall look briefly at some current approaches to this problem at the end of this chapter, insofar as they have applications to our present theme of power production.

In this chapter, more than in any other, we are nearest to current research frontiers and furthest from practical, demonstrable hardware. It is accordingly less detailed and analytical than some others and is intended mainly to provide the reader with a background in a part of the field which is changing rapidly.

9.1 Photodissociation

We have already seen, in earlier chapters, how the absorption of radiation by atoms and molecules can lead to a variety of physical effects. When we were concerned with the increase in temperature of illuminated bodies, the significant effect was that of increased motion of the molecules, either by oscillation in a solid or by translation and rotation in a liquid or a gas. These motions are themselves often the last effects of a series of distributions of the energy of the incoming photons. In examining photoelectric phenomena, we were concerned with the immediate consequences of the absorption of photons by electrons in substances with different energy band structures. Now we must consider a further set of consequences: the splitting-up or LYSIS of molecules and their re-combination into other chemical forms.

From the point of view of energy conversion, we might seek, through this process of PHOTOLYSIS, to store the energy of solar radiation in stable chemical combinations. From these, it must be recoverable at will, say by combustion. One of the first possible processes of this kind which comes to mind is the dissociation of water into its constituents, hydrogen and oxygen. If this were possible, using the energy of solar radiation, we would obtain an indefinitely-storable and transportable fuel from one of the cheapest and most abundant of raw materials. Such a process could be represented in principle by the reaction

$$2H_2O + \text{radiation} \rightarrow 2H_2 + O_2. \qquad 9.1$$

The energy required to bring this about would be recoverable (at least in part) by burning the hydrogen with the oxygen in a furnace or internal combustion engine, or by recombination in a fuel cell, giving an electric current. The consequences would be of immeasurable importance to man, so it is worth seeing whether such a process is possible or not.

We can soon see why the direct transformation represented by equation 9.1. does not occur regularly in nature. (If it did, there would be a lot of hydrogen in the air and not so much sea!) It could happen only if the energy of a single photon would be sufficient to dissociate a water molecule. The chances of a given molecule absorbing a second photon before losing the energy obtained from the first are entirely negligible. Even in the favourable case of a gas or vapour at ordinary temperatures, a molecule experiences some 10^9 collisions with its neighbours each second, so that any excess energy would be expected to be shared out quickly. Now the energy required to dissociate water into hydrogen and oxygen can be found by the process of ELECTROLYSIS. In this, familiar to many from a common school experiment, the process of charge separation is brought about by the application of an electric field and the work required is easily measured. It is found that to dissociate one water molecule requires about 3 eV of work. If this were to be provided by a radiation photon, it would have to have, according to equation 8.2, a wavelength of less than about 0·4 μm. Only about 3% of sunlight at sea level has wavelengths in this region, so that in the most favourable circumstances, such a process could not have an efficiency of more than about two per cent. Even this, however, might be worth exploiting if it could be done cheaply enough. The difficulty is, that the process cannot take place, even at an efficiency of this order, because water is nearly transparent to radiation of these wavelengths. That is to say, the mechanisms by which a water molecule can absorb a photon with wavelength around 0·4 μm, are very weak. In fact, as shown in Figure 9.1, water only begins to show significant absorption at wavelengths well below 0·4 μm. Photons in this region have enough energy to bring about the reaction shown as equation 9.1, but

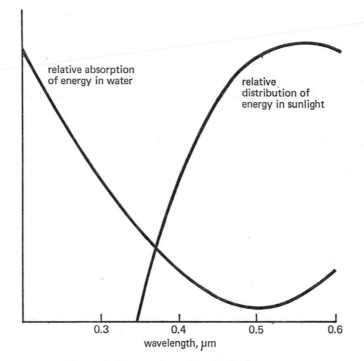

relative absorption
of energy in water

relative
distribution of
energy in sunlight

wavelength, μm

0.3   0.4   0.5   0.6

Fig. 9.1 Absorption of sunlight in water

there are no such photons present in the solar radiation at sea level.

Similar considerations apply to another very desirable reaction, in which water would be dissociated, in the presence of oxygen or oxygen-rich compounds, to form hydrogen peroxide, H_2O_2, as shown in equation 9.2:

$$2H_2O + O_2 + \text{radiation} \rightarrow 2H_2O_2. \qquad 9.2$$

The peroxide is stable and largely transparent to light. The energy stored in the reaction 9.2 could be recovered at will by causing the reverse reaction to take place. This can be accomplished merely by passing the hydrogen peroxide over crystals of potassium permanganate, which are able to act as a CATALYST for the reaction, not being consumed at all in the process. (This reaction was used in the Walter rocket engine developed in wartime Germany, the oxygen and steam produced at high temperature being employed directly as the propulsive gas). Now reaction 9.2. requires only about $1\frac{1}{2}$ eV, so that, in principle, it could take place in light of all the visible wavelengths. However, as Figure 9.1 shows, water is unable to absorb sufficiently at these wavelengths to bring this about.

Similarly, discouraging conclusions are reached for a number of other promising energy-storing reactions, though the search for better ones continues. Many effects are capable of frustrating the achievement of permanent energy storage. In some cases, the products of dissociation are too reactive and they re-combine almost immediately. In others, the products themselves absorb radiation and are thereby broken up into less useful intermediate compounds. The search for suitable reactions is necessarily a slow and painstaking activity. The original material need not be cheap and plentiful, in principle, since it is to be regenerated in the reaction in which the stored energy is later recovered, and it could thus be used again. This widens the scope of enquiry to include a fair proportion of the millions of chemical compounds known. As this work proceeds, however, studies are being made of the possibility of promoting the dissociation of water, equation 9.1, and certain others by another means, which involves the introduction of another compound in which the absorption first takes place. We can approach this method through the example of photography, in which it has long been exploited.

9.2 *Photosensitisation*

Although photography has been in use for over 100 years, and has becomes a process of very great importance, only recently has any real understanding of its photochemistry become possible, and this is still incomplete. It is interesting enough to be worth summarising briefly. The active agents in a photographic film are a collection of minute crystals of one or more salts of silver, mainly silver bromide. In such a crystal, known as the ionic type, the silver and bromine ions form a regular lattice, but it seems to be essential for the photographic process that the crystals must be formed in a way which causes each one to contain some structural irregularities. Upon exposure to light, events occur of a type we have already met; some electrons are raised into conduction bands and become free to move about in the crystal. They soon become trapped in the vicinity of the irregularities mentioned and tend to be taken up by silver ions there, which in the process become complete silver atoms. When the film is developed, a developing solution is used which is not quite powerful enough to affect an unexposed crystal, but it appears that the presence of the few complete silver atoms in an exposed crystal helps the developer to begin work and ultimately the whole crystal is thereby reduced to silver. In this way, the exposed and unexposed areas of the film are differentiated. Complicated and subtle mechanisms such as this seem to be typical of photochemical reactions.

Now it is found that the energy gap for a crystal of silver bromide is about $2\frac{1}{2}$—3 eV, so that by equation 8.2. only light with wavelengths less than about 0·45 μm can affect it. This is right at the weak deep-blue end of the spectrum, so that early films required long exposures and gave poor tonal values. It was soon found, however, that certain dyes made the film sensitive to much longer wavelengths, leading ultimately to the manufacture of PANCHROMATIC film with adequate sensitivity over the whole visible spectrum. Today, special infra-red films can be made, which have been sensitised to radiation with wavelengths out to about 1·3 μm.

Many reactions involving organic substances (compounds containing carbon atoms) can take place under illumination with ultra-violet light, in which the photons have high energy. By this means, it has been possible to synthesise substances which were not obtainable in other ways, often with high enough yields to make the process commercially attractive. Some drugs, for example, have been prepared in this way, beginning as long as 25 years ago, when the drug ascaridole (used in the eradication of parasitic worms) was produced by photosynthesis in Germany. In most reactions with organic compounds the first step seems to be the breaking of the bond between carbon atoms, the fragments with unsatisfied bonds then quickly linking up with other substances. It is known that nearly 4 eV of work is required to break the carbon bond, so that only photons with wavelengths lower than about 0·3 μm could perform this. In many cases, the photon only needs to supply the difference in energy between the bond to be broken and the new bond being formed, but many known organic changes require 3 eV or more. Yet some of these reactions have been sensitised by certain dyestuffs so as to be activated by sunlight in which the majority of photons possess less than 2 eV of energy each.

The most obvious characteristic of these PHOTOSENSITISERS is that they are densely coloured, that is, they absorb radiation strongly, with greater absorption at some part of the visible spectrum than at others. Carbocyanine dyes, for instance, widely used to sensitise photographic emulsions, have the strong turquoise colour known as CYAN; the green colour of plants comes from the photosensitive compound CHLOROPHYLL, which has also been used to sensitise numerous reactions in the laboratory.

Despite intensive study, it has to be admitted that the action of photosensitisers remains somewhat obscure, though such is the complexity of even the simplest photochemical processes that this ought not to be really surprising. Since a certain fixed amount of energy must be provided to bring about a given reaction, it is evident that the principal effect of sensitisers must be to store energy in some manner until enough has accumulated from the absorption of photons to make the reaction possible. The reaction observed is thus the result of a number of inter-mediate steps, and much recent work has been aimed at eluci-dating the sequence of events in photochemical processes. Each new technique employed in this study seems to reveal the presence of further highly reactive and short-lived fragments. With encounters between particles occurring in liquids at a rate of some 10^{10} per second, it is hard to see how the energy resulting fron one photon absorption can be retained for long enough for significant storage to take place. It is known that some kinds of this sharing-out of energy, or COLLISIONAL DEACTIVATION, take place more slowly than others. As we have seen earlier, the energy of molecules is manifested in four forms—translational, rotational, vibrational and electronic. Of these, the interchange of the translational and vibrational forms during collisions between molecules seems to take place much more slowly than others, so that a molecule excited into a high vibrational state is deactivated slowly. In effect, this means that it can store absorbed energy for some time, perhaps through a million collisions or more (though this might take only a ten-thousandth of a second!). However, the longest-lived excited state is more subtle and is now believed to be a major factor in many photosensitised reactions. This is the TRIPLET STATE. It is distinguished from others by the different SPIN of its electrons. The word *spin* is appropriate for our present model of the electron, though its interpretation is difficult for more complex models. If the electron is imagined to be a particle following an orbital path, then it can also be imagined to have rotation about its own axis, like that of the earth once a day in its motion round its orbit once a year.

For reasons which we need not consider here, the electron is found to be capable of only two distinct spin states, in which the spin is given the value $+\frac{1}{2}$ or $-\frac{1}{2}$. The spin state of the atom or molecule as a whole is represented by its MULTIPLICITY, a quantity defined, for reasons which need not detain us, as *twice the total electron spin plus one*. In the lowest energy state, or ground state, the electrons are found to be paired off so that there are equal numbers with spins of opposite sign. For this state, the total electron spin is thus zero, and the multiplicity is merely *one;* this is called the lowest SINGLET state. Now it happens in some absorption processes that the excited molecule, instead of reverting back to the ground state by collisional deactivation or

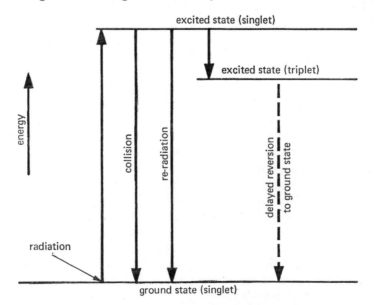

Fig. 9.2 Representation of energy changes in molecules

any other process, shifts into a state in which the spin of one of its electrons is reversed. A representation of the relation between the states is shown in Figure 9.2. The total electron spin is now $\frac{1}{2} + \frac{1}{2}$ and the multiplicity becomes *three*. This is the first TRIPLET state. It usually has only a little less energy than the original excited state and has a long lifetime. The reversion of the spin of the electron to its former value may take from a ten-thousandth of a second to as long as a second or so.

Here is a powerful method of energy storage in sensitising substances which is sufficient to explain the retention of absorbed energy in particular molecules long enough for more than one photon to influence a reaction.

9.3 Sensitisation of water photolysis

We began this chapter by proposing two ways in which water might be split up by photolysis into stable energy-storing compounds which could be used later for power generation. We have also seen that mechanisms are known whereby the energy of one photon could be retained until that from others becomes available. However, it remains a melancholy fact that no-one has so far succeeded in sensitising the photolysis of water so that it takes place in sunlight with a useful yield.

Nevertheless, steps have been taken towards this and work on it continues. A partly successful method was described as long ago as the early 1950's by L. J. HEIDT of the Massachusetts Institute of Technology, and has frequently been cited since. In this case, the photosensitisers were the perchlorates of the 'rare earth' element cerium. This element is one of a number of transition elements which can have more than one value of the valency. (The second valency value arises through the promotion of electrons from an inner shell so that they can help to form bonds).

In the following section, when we examine ionisation more fully, we shall see that when metallic salts are dissolved in a solvent such as water, the ions become separated. The metal ion which formed part of each molecule of the original salt loses some of its electrons to the radical ion, so that the latter has excess negative charge and the former a deficiency of electrons, interpreted as a positive charge. The number of charges carried is equal to the valency of the atom from which the ion is formed.

When the perchlorates of cerium are dissolved in water, for example, the CERIC and CEROUS ions are freed, the former, Ce^{4+}, needing four electrons and the latter Ce^{3+}, needing three electrons, according to the two values of the valency. A simplified representation of the Heidt reaction shows that it involves *two* photo-activated events: the simultaneous reduction of the ceric ion (acceptance of an electron) and the oxidation of the cerous ion (donation of an electron), after the manner of the following equations:—

$$4\ Ce^{4+} + 2H_2O + radn. \rightarrow 4\ Ce^{3+} + 4H^+ + O_2,$$

$$4\ Ce^{3+} + 4H_2O + radn. \rightarrow 4\ Ce^{4+} + 4OH^- + 2H_2 \quad 9.3$$

and
$$4H^+ + 4OH^- \rightarrow 4H_2O.$$

When these equations are taken together, the nett result is

$$2H_2O \rightarrow 2H_2 + O_2,$$

which we had as equation 9.1. The cerium ions have been restored to the original constitution, and four of the water molecules have recombined after a brief separation into hydrogen and hydroxl ions (H^+ and OH^-).

Unfortunately, it is found that this interesting reaction cannot be made to occur at wavelengths above 0.4 μm, so is not a practicable one for commercial exploitation. (It also suffers from other disadvantages, such as inhibition by small amounts of substances commonly present in water, and a very low yield even in ideal conditions). Although it was first reported more than 15 years ago, efforts to improve on it have not been notably successful. It will not be useful for us to attempt to review this work here. The search continues, with intensive study of the earliest stages of the process, to find a sensitiser which would be active over the whole solar spectrum, be capable of storage of energy in its excited state and mediate the photolysis of water with high efficiency in a process in which the sensitiser would be reconstituted. If found, this substance, M, say, might operate through a single photo-activated reduction step such as

$$4M + 2H_2O + radn. \rightarrow 4M^- + 4H^+ + O_2, \qquad 9.4$$

liberating oxygen, followed by an ordinary chemical oxidation

$$4M^- + 4H^+ \rightarrow 4M + 2H_2, \qquad 9.5$$

yielding hydrogen and reconstituting the sensitiser. All that can be added here is that it has not yet been shown that such a substance cannot exist.

9.4 *The photochemical cell*

One of the instances of photochemical activity cited at the beginning of this chapter was the observation by Becquerel that the potential difference across a chemical cell changed when light was allowed to fall on one of the electrodes. To understand this and to follow discussion on the possibility of energy conversion in such cells, we need to know why a potential difference occurs in the first place. This is another point at which some examination of fundamentals is desirable.

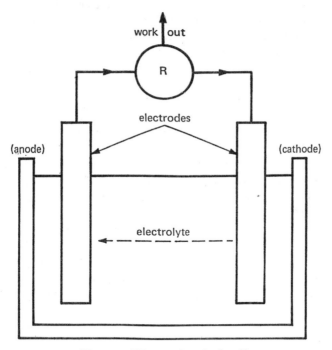

Fig. 9.3 Electrochemical cell

A chemical cell consists of two electrodes, usually metallic, immersed in a substance called an ELECTROLYTE, as shown in Figure 9.3. An electrolyte is usually a liquid or a paste, consisting of one or more compounds or a solution of acids, bases or metallic salts. It has the characteristic of conducting electricity when the electrodes are connected to a battery or other source. For this to happen, something must be able to carry electrons through the electrolyte from one electrode to the other. The nature of the carriers was a matter for conjecture for many years, though the modern ionic theory of the process is usually said to have developed from the foundations laid by the Swedish chemist ARRHENIUS in 1887. FARADAY, however, had shown in 1833 that passing current through a cell caused the deposition of metals or the evolution of gases at the surfaces of the electrodes, where the current entered and left the electrolyte. Evidently, it is here that the electrochemical reactions take place. Faraday's laws showed, in effect, that the charge carriers were fragments of the compounds of which the electrolytes were composed. Each of these fragments was called an ION (from the Greek word for *wanderer*), now known to be an atom or group of atoms (a RADICAL) having an excess or deficiency of electrons. Being charged, they were impelled to move by Coulomb forces to the electrode having opposite charge, there to be 'discharged'. This we recognise as giving up or receiving the appropriate number of electrons to make neutral atoms, which are deposited on the electrode or liberated as gases. It was subsequently realised that a multitude of other processes can take place near the electrode, all involving the transfer of electrons within the components of the electrolyte and between these and the electrode.

The occurrence of ions in electrolytes is no longer mysterious. It is evident from the manner in which they deflect X-rays that many compounds always exist in the ionic form. Common salt, for example, usually represented as the compound NaCl, consists of separate sodium and chlorine ions in a regular pattern. No actual molecule, as a discrete entity, is found in the solid state. Each of the chlorine ions, which have a negative charge, is formed by the transfer of an electron from a sodium atom, which becomes a positively charged ion. It seems that when atoms meet, they compete for electrons. The transfer of an electron from one atom to another leads to a kind of bond between them, for they thereby become charged and Coulomb forces begin to act. When such a salt is dissolved in water, however, the large separation between ions weakens these forces and the ions can

216

become completely dispersed. We then have a strong electrolyte.

This IONIC bonding is typical of many salts and bases, but most acids are formed by COVALENT bonds, in which electrons are not transferred completely but merely shared. These compounds have distinct, electrically-neutral molecules held together by these electrons, which belong to the atoms jointly. Nevertheless, many of them ionise to a greater or lesser extent in water, as a result of reactions in which the water is also ionised.

The behaviour of ions in solution is rather complex. They are charged, so they attract and repel each other, being influenced in turn by a multitude of others during their random thermal motion. Water molecules and ions affect these interactions. If the concentration of the dissolved substance is very high, the ions are close together and association between pairs of ions can be maintained even for strong electrolytes, as in their solid states. In very dilute solutions, on the other hand, virtually complete ionization can occur, even for weak electrolytes. Thus the *effective* concentration of ions in a solution, called the ACTIVITY, may be less than the actual concentration. For weak electrolytes there may be substantial numbers of unionized molecules remaining.

We can now see what happens when a metallic electrode is immersed in the electrolyte. Reactions occur, involving the atoms of the electrode and the ions of the electrolyte, in which electrons are transferred between them. The simplest example is all that we need consider here; this is when a metallic electrode is immersed in an electrolyte containing ions of the same metal. Two processes can occur. Atoms of the metal may go into solutions as ions, leaving excess electrons behind in the electrode, as indicated by the equation

$$M \rightarrow M^+ + e. \qquad 9.6$$

On the other hand, ions from the electrolyte may be deposited as atoms on the electrode, a process demanding the expression

$$M^+ + e \rightarrow M, \qquad 9.7$$

which is the reverse of equation 9.6. Both processes will continue until a state of equilibrium is reached, in which both are proceeding at the same rate. The electrode will then usually be found to have an excess or a deficiency of electrons, that is, it will

be charged. The degree of charge is called the ELECTRODE
POTENTIAL. Because the process represented by equation 9.7
is controlled by the activity of the ions in the electrolyte, the
potential developed depends upon the magnitude of the activity.

Now if two electrodes having different potentials are available, they can be connected together through an external circuit,
through which a current can flow because of the potential
difference. We then have an electrochemical cell, as in Figure
9.3. The current will flow and ions will migrate to the electrodes
until changes in concentration cause the difference of potential
between the electrodes to be reduced to zero. The cell is then
discharged. For it to provide a continuous supply of power in the
external circuit, we have to ensure that the activities of the ions
in the vicinity of the electrodes are maintained at levels which
will sustain the potential difference between the electrodes.

One of the simplest cells is that in which two identical metal
electrodes are immersed in a solution of ions of the same metal.
A potential difference can only occur in such a cell if the activity
of the ions is different in the vicinity of one electrode from its
value near the other. A difference of activity could be brought
about in weak electrolytes by illuminating one of the electrodes,
as in the cell described by Becquerel. As we have seen, radiation
can produce a variety of effects, ranging from the creation of
simple excited states to complete ejection of electrons from the
atom. Most of these effects can influence the equilibrium of the
events taking place at the electrode, symbolised by the reactions
represented by equations 9.6 and 9.7. Thus, by allowing solar
radiation to fall on one of the electrodes, the energy of the photons
can be used to drive electrons through an external circuit and to
do work there.

By now, the reader will expect to learn that no reaction has
yet been found in which this process occurs with a high enough
efficiency to be commercially attractive. Nevertheless, a number
of possible reactions have been shown to work in favourable
conditions—principally under illumination by ultra-violet light,
for which the photons have high energy. The efficiency of a
PHOTOCHEMICAL CELL of the type we have considered depends
primarily upon three factors. Firstly, there is the efficiency with
which the light energy is used in the primary absorption process.
We have seen already that the nature of quantum reactions sets

a maximum value of about 45% on this process, because of the distribution of the Sun's energy with respect to wavelength. Secondly, the overall efficiency is directly affected by the relative speed of the reverse process, or BACK-REACTION, compared with the rate at which ions can migrate to the electrode surface, which defines the ion MOBILITY. Lastly, there are changes in the electrode reactions which occur when a current is being drawn. One of the most serious is the development of OVERVOLTAGE, in which the electrode potential varies with the current density. This too is due largely to limitations on the mobility of ions, which can move only at a moderate rate towards the electrodes because they are impeded by interactions with other ions. Just as with the photoelectric cell, the potential difference falls from its maximum on open-circuit to zero at short-circuit, a best operating condition having to be found between these extremes.

There are millions of potentially suitable electrode reactions for cells, if organic as well as inorganic ions are considered. For most, the *rate* at which the reactions and back-reactions can proceed and the associated factors such as the ion mobilities cannot be predicted accurately in the present state of knowledge. The great enthusiasm of recent decades for the possibility of developing a power-producing cell run by solar radiation has given way to caution in the face of the task of investigating all possible promising reactions. At the present time, the caution remains optimistic, though the highest predicted overall efficiencies for reactions proposed so far are very disappointing.

9.5 *Photosynthesis*

When it has proved so difficult to devise processes by which the energy of the sun's radiation is converted or stored for later use, it is natural to turn for guidance to the method used by living cells, which can do these things with remarkable economy. Under good conditions plants have the ability to store, in the form of combustible matter consisting of large molecules, some 10% of the solar radiation falling on them. The world total of energy conversion by this means is staggering; one estimate puts it at ten times the present rate of energy consumption by mankind for the plant life on land alone. The minute surface-living organisms that teem in the seas must use much more.

The engineers' ultimate aim in this field must be to imitate this process or some part of it in a controlled manner. It has, however, proved to be an extraordinarily complex process, and although many of the steps have been elucidated, much remains incompletely understood. Whilst study continues, therefore, attention is given to the direct use of plants as energy converters in an otherwise man-made system. In this section, we shall look at a few of the salient features of photo-synthesis in plants as a preparation for a consideration of this process as part of an energy producing system.

The overall photo-activated reaction which takes place in all green plants is often represented by the equation

$$nCO_2 + nH_2O + radn. \rightarrow n(CH_2O) + nO_2 \qquad 9.8$$

This is intended to show that in the plant the two essential reactants, carbon dioxide and water, are combined into CARBOHYDRATES, with the liberation of oxygen. Whilst this reflects the essential facts about plant metabolism, it is, of course, a gross simplification. Carbon dioxide and water must certainly be provided (the former is usually taken from the air, in which it is present at a concentration of about 0·03%). Oxygen is certainly liberated, at a rate of one molecule per molecule of CO_2 inspired, as the equation shows. The carbohydrate molecules, however, are subject to a chain of chemical transformations through which the plant builds the organic substances needed for its growing structure. The simplest carbohydrate of the form $n(CH_2O)$ is GLUCOSE, $C_6H_{12}O_6$,—in which n = 6. This is present in some degree in all plants, but for the most part, it is a building block for the more elaborate molecules, the fats, proteins and higher carbohydrates present in such profusion in the living plant. We need not pause to consider the processes used in these subsequent manipulations, except to note that they seem to be brought about by the catalytic action of very complex protein molecules known as ENZYMES, a specific enyzme being used for each step. The energy required for these building operations is derived from the oxidation of some of the early carbohydrates back into CO_2 and water. It is these subsequent building operations that depend for their efficiency on tiny amounts of trace elements in the soil, the provision of which sometimes makes the difference between bountiful cropping and famine.

Paths taken by the reactants in equation 9.8. have been studied by using atoms which have been 'labelled' by making them radioactive. This revealed that the oxygen liberated in the process comes exlusively from the water, and none from the carbon dioxide. Thus, the core of the process is the lysis of water molecules, just as in the first equation of this chapter, equation 9.1. We saw in earlier sections that the photons of visible light have individually insufficient energy to perform this act and that a sensitiser is needed, which can absorb photons and store their energy until enough has been accumulated. In green plants this sensitiser is a magnesium porphyrin known as CHLOROPHYLL-A, having the composition $MgN_4C_{55}H_{72}O_5$. This green substance is densely concentrated in the CHLORO-PLASTS of plants. It absorbs light throughout the visible spectrum, though most strongly in the red, the deficiency of red in the reflected radiation giving it its pronounced green colour. It is believed to carry out its catalytic functions through excitation into one or more of its triplet states.

The energy which has to be stored is substantial. Including that which is required for the reduction of carbon dioxide, it amounts in all to about 5 eV per carbohydrate unit synthesised. Yet the process can be driven by light with wavelengths right out to the red end of the visible spectrum, where the energy of a photon is only about 1·7 eV. These photons, however, do not seem to be used with perfect efficiency, though the precise utilisation is still a matter of debate. Most observers agree that about *eight* photons of red light are required per carbohydrate unit synthesised. If so, the energy taken up would be about 14 eV. Since the energy stored is about 5 eV, the overall efficiency of energy storage in red light is thus about 35%. It is supposed that the losses accumulate from the multitude of steps in the process of synthesis, the majority of which are therefore assumed to be irreversible in some degree.

In Section 8.2, we saw how the overall efficiency of utilisation of the sun's energy could be calculated for a simple quantum process. If the threshold for photosynthesis occurs at the extreme red end of the spectrum, (say 0·7 μm) the fraction of the energy lying beyond it in the infra-red (about 55%) cannot be used. Moreover, if the process is similar to that involved in photoelectricity, the shorter wavelengths cannot be used effectively, either. If we suppose for simplicity that we can apply to photosynthesis the method of Section 8.2, with a 35% efficiency for light near the cut-off wavelength, we find that the maximum overall efficiency of utilisation of solar energy by plants would be about 11%. Values approaching this are obtained in closely-controlled experiments, though for plants growing freely in the open, the efficiency is often an order of magnitude less, say about 1%. It is this potential margin for improvement that spurs on the plant geneticist and soil scientist and gives us hope that the earth may yet be made to feed us all in the next critical decades.

9.6 Plant culture for energy storage

We are mainly concerned, in this book, with the production of power from solar radiation. To achieve this through the growth of plants, we must recover the energy stored in organic material by burning or some other disintegrative process. When plant material is burned in air, it yields about 6 kWh of energy in the form of heat per kg of material burned. In the most favourable cases, such as the culture of sugar cane in moist tropical regions, where growth can continue all the year round, the annual yield of dry organic matter can be as great as 10 kg per m² of ground surface. Burning this would yield about 60 kWh/m² for an area in which the annual insolation (see Chapter 2) is about 2 000 kWh/m². Thus we could recover about 3% of the sun's energy in this way.

Many factors prevent the achievement of the maximum theoretical efficiency in plants grown conventionally. Some, such as the shading of one leaf by others and the loss represented by spaces between plants, are in principle open to adjustment. Others are more fundamental. One of these is the adequacy of the carbon dioxide supply. In the open, CO_2 reaches the leaves of plants through mixing processes in the air. These processes are determined by the structure of eddies in the flow and are accelerated by winds. In still conditions, mixing occurs only by diffusion and thermal convection currents and this might be inadequate. The demands of rapidly-growing plants are prodigious. To produce 10 kg/m² of dry matter per year requires almost the same amount of CO_2 per year, corresponding to an average of about 0·015 m³ per day for each square metre of surface. At the peak of the growing period, the requirement would be as much as 0.04 m³/m² per day. At an atmospheric concentration of 0.03%, this is equivalent to all the CO_2 contained in the air above the crop up to a height of some 130 metres. Faced with these figures, it is hard to believe that the supply of CO_2 is always adequate for the quickest-growing plants.

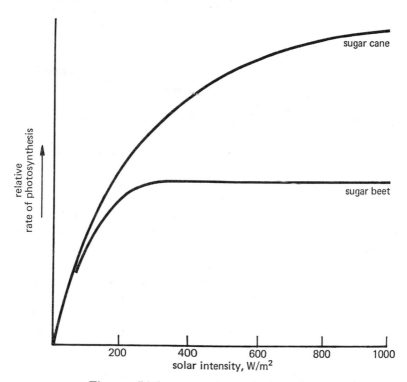

Fig. 9.4 Light saturation in higher plants

Even when the CO_2 supply is made ample under experimental conditions, however, another, more serious limitation arises. It is found that the rate of photosynthesis is not proportional to the rate of incidence of radiant energy. Typical characteristics are shown in Figure 9.4. In some plants, the rate of photosynthesis ceases to increase beyond a certain level of illumination. This 'light saturation' is thought to be due to one or more slow steps in the synthesis processes, so that at high intensities, photons are arriving at the chlorophyll molecules faster than their energy can be used further down the chain. Since this is more marked in some species than in others, it is to be expected that plants having the best characteristics for maximum photosynthesis in a given region could be developed by suitable selection and breeding.

The rate of synthesis in plants is also markedly affected by temperature and humidity. Those plants which are well adapted to strong sunshine conditions show a steady increase in growth rate up to temperatures of more than 30°C, but this is inhibited if the air is very dry. CO_2 is absorbed from the air through openings in the leaves—the STOMATA. In low humidity conditions, large quantities of water vapour can escape through these openings and to prevent dehydration and wilting of the plant, a response sets in which leads to the closing of the stomata. This in turn limits the rate of CO_2 uptake and hence the rate of growth. The highest growth rates are thus usually found in wet tropical conditions, where the production of matter can reach 40 grammes per day per m^2 of exposed surface.

Yields such as this are the result of intensive husbandry with experience stretching over decades or even centuries. We have seen that the maximum theoretical efficiency for photosynthesis corresponds to yields at least some 3 to 4 times as great as those achieved so far. It is not yet clear how far this margin can be reduced, though it is thought that by breeding of plants specifically for the maximum production of combustible matter and by the most intensive cultivation, yields of the order of 20 kg/m^2 per annum should be readily realisable in the short term. This would represent about 6% trapping of solar energy. The losses in combustion, steam raising, conversion to mechanical and then electrical power, would, even using the most advanced techniques for this purpose, result in a recovery in the form of work of about 2% of the original solar energy. It remains to be shown whether such a disappointing result would justify the investment involved.

9.7 *Culture of lower plants*

Some serious attempts have been made, notably in Japan and the U.S.A., to produce organic matter on a large scale under conditions which can be more closely controlled than in the cultivation of higher plants in the open. Most of the work has been done with minute single-celled plants of the order known as ALGAE. It is these organisms whose astonishing proliferation in ponds, even in Britain, is responsible for turning the water a brilliant opaque green in spells of strong sunshine. The common species *chlorella* and *scenedesmus* are found in many varieties in open water. A typical individual organism is an oval single cell, about 10 μm long. When growing in conditions in which ample CO_2 and suitable nutrients are provided, each cell undergoes MITOSIS, or division into two, or sometimes four, daughter cells, in a period which can be as short as 12 hours, say one day of illumination. The growth of the population by 'compound interest' is such that a single cell would by this means become ten thousand million cells in a month. All strains of algae so far investigated suffer light saturation to some degree, the majority reaching saturation at intensities of about 50 to 100W/m². As a result of the influence of the slow stages in photo-synthesis, which can take place in the dark, the effects of light saturation in these organisms can be offset by alternating periods of exposure to high light intensities with periods at low intensities. In the 'dark' periods, reactions continue, powered by energy stored during the 'light' periods. Reduction of the average intensity of illumination in this manner is possible for cells which live in water by agitating the suspension so that individual algal cells pass intermittently through the bright regions near the surface and then into darker regions in the body of the culture.

Algae growing in this way have to be provided with nutrient salts, like other plants, which normally obtain them from the soil. It is also necessary to bubble CO_2 or air enriched with it through the culture to ensure adequate access to this chemical. The major plant nutrients, (nitrates, phosphates and sulphates and the small but essential quantities of trace elements such as iron and manganese) have to be provided at the same rate as they are used by the plants if continuous culture is to be possible. Some of the largest-scale experiments with algal culture have met this difficulty by the continuous supply of nutrients in the liquid extract from municipal sewage. Since the waste discharged into domestic sewers consists of organic material originally consumed as food, this represents a satisfactory nutrient for plant life, being only marginally depleted during digestion. Its use for algal culture is comparable with its use as a fertiliser for crops, a practice already used extensively in some areas.

All sewage contains BACTERIA, and a co-existence between bacteria and algae ensues which is beneficial to both. Bacteria are also simple micro-organisms, somewhat smaller than the unicellular algae. Many varieties live in the ducts and passages of humans and animals and are discharged in great numbers in their faeces. Countless other varieties inhabit the soil, water and the air. Their principal functions in ecology are to provide nitrogenous material in forms which can be utilised by plants and animals, by capturing or FIXING atmospheric nitrogen, or by breaking down into simpler compounds the complex organic molecules such as are present in the residue of plants and the faeces of animals. The types of bacteria operating in conjunction with algal growth are the oxygen-requiring, or AEROBIC varieties. They break down the organic matter in sewage into substances which the algae need for growth—CO_2, water and nitrates, phosphates and other nutrients. In turn, the algae respire the oxygen needed by the bacteria. Colonies co-existing in this way have lived continuously in outdoor tanks throughout the year in several parts of the world. Vigorous growth has been reported even for outdoor cultures in northern Europe.

The continuous growth of algae along these lines, with the attendant decomposition of wastes and the production of water largely cleared of distasteful or dangerous matter, is an attractive proposition. In effect, it is the bringing about, under controlled conditions, of the decomposition and resynthesis of organic matter which otherwise occurs naturally. We cannot conveniently represent here in simple terms the equations governing the steady state of such a system, depending as it does on the concentration and nature of the waste stock, the activity of bacteria and algae, and the influence of solar intensity, temperature and other factors. Enough large-scale experiments have been done, however, to show the potentialities of such a scheme. The algae can be regularly harvested, to maintain a roughly constant concentration, by continuously passing some of the culture medium through a CENTRIFUGE (operating like a domestic spin drier). When the energy potentially available in the dried algae is related to the solar energy incident upon the growth tanks, it has been shown that up to 8% can be recovered—approaching quite closely the theoretical maximum deduced in Section 9.5. This represents a yield of dry matter at a rate of about 30 kg per m² per year, though the yield when operating continuously is usually closer to the annual yield of the higher plants, say 10–15 kg/m² year.

9.8 *Energy recovery from plants*

Most of the experiments in the culture of algae, like many trials in plant culture, have been aimed at producing nutritious foodstuffs. Algae have much to recommend them as foods. They have higher percentage contents of protein, fats and vitamins than any other vegetable source, excepting a few seeds such as the soya bean. This is, perhaps, because their small size and water-borne existence has not required them to convert much of their carbohydrates into cellulose, the woody fibre which forms the skeleton of higher plants but which cannot be digested by animals. Tests have shown dried algae to be nutritionally comparable to yeasts and to some animal protein sources such as skimmed milk and egg white. It is obvious why they have been examined as potential new sources of protein, of which the world's deficiency is becoming grave. However, we are concerned here with their potential as a fuel, and in this they are not materially more attractive than the higher plants.

One of the disadvantages of algal culture is that the organisms have to be separated out of a weak suspension in water, and this requires energy. The separated cells, even at this stage, contain about 75% water within themselves so that for storage or transportation they would need to be dried further. Again, energy is required; even if solar drying is employed, potential power generating capacity is being used up. The dried algal cells are fragments of matter about 1 μm in diameter. Although there has been no requirement in the past for the development of the necessary technology for the combustion of such a fine powder, it would not appear to present much difficulty. Fed as a suspension in the air necessary for its combustion, it could be used to fire boilers or directly as the fuel for gas turbines or even reciprocating heat engines.

An alternative to the drying of algae is to decompose, or

DIGEST, them in closed chambers by the action of other varieties of bacteria which live without oxygen; the ANAEROBIC bacteria. This is a process widely used in sewage disposal and for which the technology is well established. The slurry of partly-decomposed sewage and algae from the growth tanks is fed into closed digesters, in which the organic matter is broken down almost completely, yielding water enriched with nitrates and phosphates (used for fertilisers) and a combustible gas. Existing sewage plants are not uncommonly powered by burning this gas, which contains more than 60% of methane, CH_4, the simplest member of the paraffin series of hydrocarbons. Although its boiling point is much lower than that of its relative, propane, C_3H_8 (widely used as a fuel) it can be liquified, stored and transported readily in refrigerated vessels. A possible arrangement for a power plant using this procedure is shown schematically in Figure 9.5.

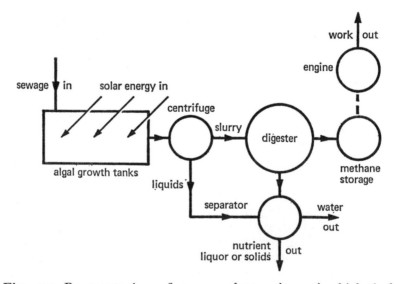

Fig. 9.5 Representation of power plant using microbiological systems

Conversion to methane in this manner is considered to be the simplest and cheapest method of using algal growths as a fuel. A possible alternative process is the conversion of the carbohydrates by yeasts into alcohol, represented by GAY-LUSSAC's equation:

$$C_6H_{12}O_6 \rightarrow 2C_2H_5OH + 2CO_2. \qquad 9.9$$

These methods are, of course, equally suitable for the conversion of matter from higher plants and, indeed, are some of the very few possible ways of recovering the energy stored in this material. The overall efficiency of conversion of solar energy via plant matter to a form such as electrical energy is, however, very low when compared with that of some other methods discussed in previous chapters. When referred to the exposed areas of the culture tanks from which the algae were drawn, or of the fields supplying higher plant material, the power production by such a chain would be unlikely to exceed about 70 Wh per m² per day. The possibility of a useful place for this and other systems in a world badly in need of new sources of power is the subject of the next, and final chapter.

10

THE INTRODUCTION OF SOLAR POWER

A world where nothing is had for nothing.

ARTHUR HUGH CLOUGH (1819–1861)

B ECAUSE it pours down, nearly everywhere and without effort on our part, solar energy is often considered to be free. Many have been caused to wonder, therefore, why this vast source of energy has not long ago been made to supply us with all the power we need at negligible cost. But the sun's energy is no more free than other sources of energy. All are there for the taking, but the taking costs money, and the taking of solar energy costs more than most. In this chapter, we shall look briefly at some of the features of solar energy which make it expensive to recover and draw together the various methods of recovery to see what are the prospects for each.

10.1 *Intensity of solar energy*
One of the obstacles to universal exploitation of solar energy is its low intensity. We saw in Chapter 2 how to calculate the intensity on any surface at any latitude and any time. Even for the very clear conditions to which that chapter referred, it was plain that the low intensity (and the variability) of the energy were grave disadvantages in a potential source of power.

At noon in the tropics, the intensity may approach 1 kW per m² of exposed area. Even the best of the many devices considered earlier for converting solar energy to more convenient forms would not deliver more than about 150 W/m² even under these ideal conditions. The daily output would only be about ½ to 1 kWh/m². Evidently, the size of the collectors alone would make it uneconomic to supply in this way anything other than small local power demands. In the most developed countries, the total energy demand (by the definition of Section 1.3), is already running at about 50 kWh per person per day. To meet such a requirement for a moderate-size community of say

100 000 people (roughly equivalent to a British city the size of Gloucester or Cambridge) by the most efficient method presently available, collectors with a total area exceeding 5 km² would be needed. This approaches the area of the ground on which the city is built, even at the moderate density of older urban development. If a zone around the edge of such a city were used for collectors, it would need to be hundreds of metres wide. Where the climate is less favourable, there will be many parts of the world where the rate of energy consumption exceeds that which arrives from the sun on the entire land surface area, inhabited or not.

Schemes have been proposed for siting solar collectors along highways and railways and, indeed, in any place where they would not occlude the light from precious agricultural land. It can be shown that significant contributions to the energy supply could be provided by such means in developed areas. Nevertheless, although we cannot say that the time will never come when such schemes would be acceptable, it is safe to conclude that for the forseeable future the communities of these areas will continue to obtain their energy in other ways.

For the more modest demands of a developing or rural community, however, the outlook is very different. Here, the provision of energy, at a rate much less than that at which solar energy falls on the region, would make a dramatic change in the fortunes of the area. Moreover, sun-powered devices could well be competitive with other sources. There is a lesson here in the success of the solar water heater. As we saw in Chapter 5, the hot-water demands, even of developed countries, can be met by devices whose collector area is less than the roof area of ordinary dwelling houses. The millions of these devices already in use testify to their competitiveness in these conditions.

There are countless small tasks, at present performed by manual or animal labour or not at all, whose power demand is within the compass of solar-operated systems with collector areas from a few square metres up to a few hundred square metres. As with any other innovation, the introduction of such devices is largely a matter of economics, a notoriously-difficult topic, which we shall try to examine briefly later. In the meantime, however, we should look again at a further inhibiting factor, which applies even where the noon-day intensity is as high as it can be. This is the variation of the intensity with hour and season, and the large part of every day when there is no energy available at all.

10.2 *The variability of solar energy*

Variability of the incoming energy during the day is a feature which has to be accepted wherever solar energy is to be utilised. The methods of Chapter 2 allow us to calculate the variation during the day for clear conditions and to find the period throughout which no energy is available. There are some duties in which the corresponding variation in output of the energy converter can be tolerated and the energy used whenever it is to hand. The pumping of water for irrigation is such a duty, in which irregularity is not a serious handicap and the greatest output coincides with the greatest need. Such cases are rare, and whenever a fairly steady demand has to be met, energy must somehow be stored when it is plentiful for recovery when it is not. Some examples of ways of doing this were considered briefly in Chapter 5.

We saw in Chapter 2 that the increase in day length with increasing latitude largely compensates in summer for the effect of reduced solar altitude, so that the daily insolation in that season is much the same over a wide range of latitude. Sunshine records show, however, that considerable variations in summer insolation occur, even between points at the same latitude. This is largely due to another element in the variability, the degree of cloud cover. Although certain locations, particularly high ground immediately inland of coastlines, are known to be subject to more cloud than others, the effects are hard to predict because of wide variation in cloud type and thickness and the difficulty of defining the extent of cloud cover. Photographs from satellites are beginning to show the way in which cloud cover at any location is related to the movements of very large air masses, often covering a major part of the earth. But the variations of solar radiation in any locality depend upon the fine detail of the cloud structure at that point, which at best will only be predictable on a statistical basis after years of continuous recording there.

As might be inferred from the absorption characteristics of water, given in Figure 9.1, clouds do not absorb much of the solar energy (they would soon be dispersed through evaporation if they did). The fine droplets of which they are composed scatter some of the radiation which falls on the cloud; the rest is transmitted without encountering a droplet. Some of the scattered radiation is deflected into a generally upward direction—that is, it is 'reflected'—but much of it emerges with the directly transmitted radiation at the bottom of the cloud. The total amount emerging in this way depends upon the depth of the cloud and the size of its droplets, and it may be nearly all of the incident radiation or as little as 10% of it. For this reason, there is no simple relationship between degree of cloud cover, c, as conventionally estimated by observation from the ground, and the reduction in insolation produced by it. Many relationships have been proposed to express the dependence of insolation on simple, observable features of the local weather, with only moderate success. One of the simplest presentations, adequate for our present purposes, is that shown in terms of the cloud cover only, in Figure 10.1. In this, c is simply the fraction of the sky covered by cloud at any instant, without regard to thickness or type. The mean curve, which may be used for crude estimates, is closely represented by the equation

$$f = 1 - \tfrac{1}{2} c^2. \qquad\qquad 10.1$$

Because much of the radiation scattered by clouds joins the directly-transmitted part, the reduction of insolation remains small up to quite large amounts of cloud cover. Even for complete cover, about half the incident radiation is transmitted, on the average.

Nevertheless, the nett effect of a cloudy environment is substantial. Even when the insolation is averaged over a month at a time, the differences observed from place to place and year to year are sometimes quite large. For instance, Figure 10.2 shows some estimates of the highest and lowest monthly figures for locations with a tropical latitude of $23\tfrac{1}{2}°$. In view of such variations, which can amount to 50% or more, data must be obtained by observation over long periods at the site of any proposed installation before the design of a system using solar energy canbe started. The collection of data such as these will have to form part of the long-term market surveys which must precede the economic use of solar-powered systems on a large scale.

233

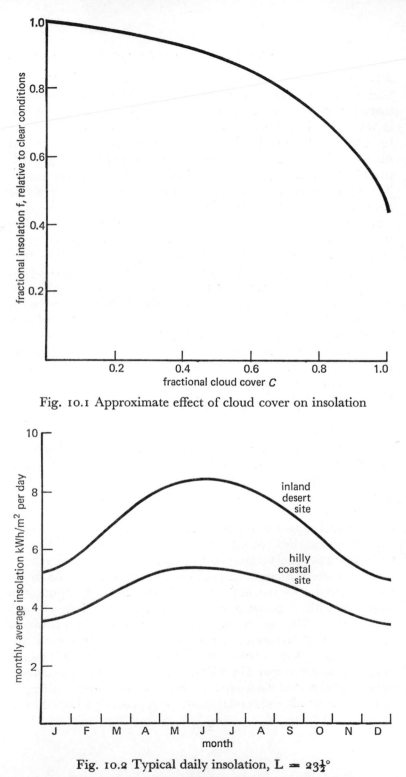

Fig. 10.1 Approximate effect of cloud cover on insolation

Fig. 10.2 Typical daily insolation, L = $23\frac{1}{2}°$

10.3 *Energy storage requirements*

Facilities for storing energy must be provided whenever equipment is being designed for heating or cooling buildings, power production for continuous use and any other purpose in which variation in the demand does not coincide with variation in the supply. In continuously-clear conditions, which obtain for large parts of the year in some desert areas, it might be necessary to provide storage equivalent to only one day's supply, since the next day's can be confidently expected. Such a low storage might still be sufficient where it is planned to use some auxiliary source whenever the supply fails to meet the demand. Most of the house heating systems designed for solar operation incorporate auxiliary heating of some kind, since it is uneconomical to provide a solar heating system to meet the requirements of the dullest and coldest day of the year and then have a plant which is much too large at all other times. There is evidently a relation in this case between the variability of the solar radiation, the variability of the heating demand, the capacity of the auxiliary supply and the amount of storage capacity provided. The balance drawn between these is primarily an economic one. We cannot explore it very far here; it depends on factors such as the local cost of fuel, material, plant and labour and the variation of outside air temperature and wind strength, throughout the year. Some of these, like the insolation, cannot be predicted reliably for a given location, but must be recorded there over a long period of time. However, it will be seen that the relation between these quantities will not be affected so much by the individual highest or lowest recorded values of insolation, as by the length of time for which the high or low values endure. One way in which solar radiation data have been presented for this purpose is as shown in Figure 10.3. This gives, for locations at about the same latitude, the number of *consecutive* days in a given season on which the observed radiation was below any selected value.

With data such as these to hand, the engineer can begin to balance the various parts of the system: the energy converter, the auxiliary plant (if any) and the storage system. The latter may take a variety of forms, some of which have been mentioned already. In Section 4.6 and 5.4 we looked briefly at methods of storage, when the energy is available simply as heat.

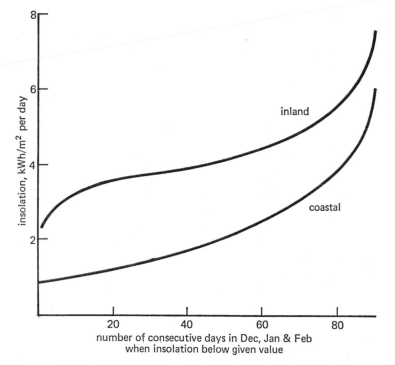

Fig. 10.3 Typical distribution of periods of low insolation, L = 40°

Then, it can conveniently be stored as the internal energy of a quantity of some suitable substance whose temperature has been raised. It is found that if a change of physical state, such as melting, takes place at an appropriate temperature, the work required is often very large and energy can be stored at rates of the order of 50 Wh per kg of substance. Nevertheless, large storage chambers would be required for quite a modest storage of a few MWh. In such a system, there is also a continuous loss of energy from the store by heat transfer to the surroundings.

When the output of the solar converter is available as mechanical work one of the most attractive methods of storage is to use the work to drive pumps which raise a quantity of water to a high reservoir. When required, the water may then be drawn off into irrigation systems or used to drive turbines from which the energy is recovered again as work. Several of these 'pump storage' schemes have already been employed in different parts of the world to store surplus energy, particularly from nuclear reactors. These reactors and many other devices work most efficiently when operating continuously at their design output. When this exceeds the demand, for example during the night, it is economical to store the surplus for later recovery when the demand is heavy. The energy storage capacity of a reservoir is large only because the volume of water retained in it is large. When a kilogram of water is raised one metre against the pull of the earth's gravity, the work that has to be done is nearly 10J, so that 3600 kg raised 100 m would require only 1 kWh of work to be expended. This can be recovered when the water is lowered again. Of course, the process is not entirely reversible; neither the pump which raises the water, nor the turbine it drives on the return journey (which can be the same machine) are perfectly efficient, and there are friction losses in the pipes for flow in both directions. Nevertheless, it is often possible to design for the recovery of 60% of the work input, though occasionally it might not be as high as this—where there are long pipe runs for example. Storage reservoirs are not cheap to construct, and require the reasonable proximity of an elevated region in which to build them. Moreover, like all reservoirs, they suffer a continuous loss of water by evaporation. In a dry climate, especially if there is wind to remove and replace vapour-laden air from above the water surface, this loss can amount to the equivalent of a fall in level of several centimetres per day. If the water has been raised 100 m, this would mean that about one kWh of work must be done each day for each 100 square metres of reservoir surface, simply to make up the evaporation loss. In these conditions, it is questionable whether so precious a commodity as water could be used for energy storage alone.

Pump storage is also a possible method to be considered when the output of the solar converter is electricity. In such a case, the converter could be arranged to drive a motor coupled to a pump, which works in reverse as a turbine and generator. The cost of these machines is a disadvantage, but this system has to be compared with other methods of storage for an electrical output. The only direct storage available in such a case is in electrochemical cells or accumulators, where the input is used to separate the charges in a chemical system and recovered on re-uniting them. The efficiency is high, but the size and cost of such cells is quite prohibitive for the storage of more than a few kWh. This can be seen through the familiar example of the battery for a typical domestic saloon car, costing in Britain the equivalent of about $20, but storing, at full charge, less than $\frac{1}{2}$ kWh of energy.

Another possibility which has been investigated is the use of electricity to bring about the lysis of water into hydrogen and oxygen. These gases are separated in the process and can be collected and stored indefinitely, to be recombined when required in a heat engine or fuel cell. With the latter, the recovery rate could be as high as 60%, and there are no losses to be allowed for during the storage process. When we discussed the lysis of water in the previous chapter, we noted that about 3 eV of work are required to separate one water molecule into its elements, a process which liberates one hydrogen molecule. Since there are $2 \cdot 3 \times 10^{25}$ eV to one kWh, we find that a perfect electrolysis system would produce about $7 \cdot 5 \times 10^{24}$ molecules of hydrogen per kWh of electricity used. At ordinary atmospheric pressure, these would occupy a volume of about $\frac{1}{4}$ m^3, so appreciable compression would be necessary if hydrogen were to be stored in an amount equivalent to some MWh of energy. Thus the capital cost of pumps and storage vessels becomes an important factor in such an arrangement. On the other hand, it would probably not be necessary even to collect the oxygen liberated in electrolysis, since this can be recovered from the air when recombination of hydrogen and oxygen is to take place.

In the solar pond, briefly described in Section 4.6, energy is both collected and stored in the same unit. This forms an attractive system for power production on a fairly large scale, with a cost per unit of power which is almost independent of size. It is, however, a low-temperature system, so that, for reasons discussed in Chapter 6, it is limited thermodynamically to a low overall efficiency.

The only other systems, among those we have considered in this book, in which the processes of energy conversion and of storage occur at the same time, are the photobiological ones. In the production of algae, for instance, the dried product is indefinitely storable and can be burned at will. Some, though not all, of the most rapidly-growing plants yield matter which can be dried and stored. All organic matter can also be decomposed by bacteria to give combustible gases which are storable. It should not be thought, however, that no cost is involved in this, for even when the drying plant is solar powered, or the digester operates with free bacteria, these parts of the plant represent a significant capital investment. One of the most attractive aspects of photobiological systems, on the other hand, the consequences of which have not yet been much explored, is the large short-term storage of energy which takes place during the growing period. Harvesting of algae might take place continuously, at a rate which represents only a fraction of the total stock every day. With higher plants, the daily harvest rate would be minute compared with the mass of the growing stock. These are conditions in which there can be great short-term differences between the supply and the demand, without disturbing the long-term stability of the system as a whole. For no other system can it be said that there is unlikely to be a significant difference between the costs of operating with a steady or an unsteady demand or supply.

10.4 *Economic factors*
WILSON's first solar-heated distillation plant began operating in Chile in 1883, followed a year later by BERNIÈRE AND MOUCHOT's solar-powered steam engine. Almost a century afterwards, we are only just beginning to see the widespread adoption of the simplest solar-operated device, the water heater. Other sun-powered systems continue to remain the subjects of cautious experiments. The reasons for this long apprenticeship are not hard to find in general terms. It has coincided with the period of extensive usage of fossil fuels and economic factors have favoured these for most of this period. Only now is it beginning to be found, here and there, that solar-powered systems can be competitive.

We have seen in earlier chapters how to calculate the performance of a variety of different systems. Each of these has been constructed and operated on an experimental basis, so that their performance can be said to contain few further uncertainties. Moreover, there have never been clear grounds for dismissing solar energy conversion on a count of low overall efficiency. Some of the devices described here could not have been built until recently, but simple heat engines operating with flat-plate collectors could have been made during most of the last 100 years. At any point in that time they would have operated with efficiencies not much worse than those of contomporary steam engines (though very low in absolute terms). Today, we might expect, on the basis of the simple analyses of this book, efficiencies of the order shown in Table 10.1.

TABLE 10.1

Approximate efficiency of solar energy conversion to mechanical or electrical work by various devices

Heat engines with flat-plate collectors	3– 5%
Heat engines with concentrating collectors	15–25%
Thermionic generators with concentrators	20–30%
Thermoelectric generators with flat-plate collectors	$\frac{1}{2}$– 1%
Thermoelectric generators with concentrators	3– 5%
Photoelectric generators	3–16%
Photobiological systems using higher plants	1– 2%
Photobiological systems using micro-organisms	2– 3%

The wide variation in these efficiencies does not necessarily mean that one system is to be preferred to all the others. Since the energy input from the sun costs nothing in itself, the capital cost of the equipment is a critical factor and this can be expected to vary widely among the systems cited. Precise comparisons cannot be made here, for they depend in turn upon highly variable factors with values specific to the proposed application and the parts of the world in which the systems are to be made and operated. The beneficial effect of volume production on unit costs will also be different for the different devices so that their relative costs will change according to the demand.

Certain features are fairly clear, however. Most of those devices with high conversion efficiency rely upon concentrating mirrors and these need to be steerable so as to follow the apparent motion of the sun. These mirrors and their steering gear are very costly and may represent up to $\frac{3}{4}$ of the overall cost of the equipment. It is not generally reckoned that effective mirror systems can be produced, even on a large scale, for less than about $200 per square metre of collecting area, for diameters of a few metres. Larger collectors cost more per unit area so that doubling the diameter increases the cost per unit area by some 30%.

These factors are strongly to the disadvantage of those devices needing concentrators, for the capital cost of flat-plate collectors is less than a tenth as much and is virtually independent of size. The differences in cost arise mainly from the much greater fidelity of shape required for concentrators and the requirements for steering and resistance to wind forces. As a result, the higher cost of devices using concentrators generally outweighs their greater efficiency, so that the flat-plate devices are able to produce power more cheaply. The capital cost of a heat-engine/flat-plate collector system is thus about the cheapest among those listed, though even in quantity production it is unlikely to be less than about $1 000 per kW of capacity.

It is not possible to obtain reliable estimates of the probable cost of thermionic, thermoelectric and photoelectric systems at the present time. Those of the latter produced for space vehicles have cost some $100 000 to $400 000 per kW capacity, but this can hardly be representative of large-scale production with much lower tolerances. Moreover, the possible place of thin-film and polycrystalline photocells is still very uncertain and might bring costs down below this by factors of hundreds. Another promising contender is the thermoelectric generator with flat-plate collector. This would become competitive with the development of low-temperature thermoelectric elements having figures-of-merit only a little above those already available. Many believe that this is one of the most likely means of obtaining quite large-scale conversion of solar energy at moderate cost in the foreseeable future.

The question of what can be regarded as a moderate cost can be answered only after exhaustive economic analysis for particular circumstances. This is beyond our present scope, though we can usefully review some of the factors involved. One factor which might be expected to set the various methods of power production in an order of merit would be the cost of power production relative to the cost of producing it by some competing system. A difficulty about this is that the cheapest competing system is not the same in all parts of the world. In one country it will be the diesel-driven generator, in another a coal-fired steam turbine plant, in a third a hydroelectric converter. Moreover, the cost of a given solar-powered system will vary widely from place to place according to the price of materials and labour, the ruling interest rates charged on borrowed money, and so on.

We have said that a heat engine system working from a flat-plate collector costs about $1000 per kW of installed capacity. There is very little technical difficulty in making plant of this kind and there are few places throughout the world where it could not be produced locally, but the price is unlikely to be much lower anywhere. Such a cost appears at first to be quite uncompetitive compared with that of large-scale sources such as conventional thermal power stations and hydroelectric plant, which is nearer to $100 per kW installed capacity. Even a small diesel-powered generator costs no more than this. However, when fuel costs are taken into account, the comparison is not so unfavourable. Low-temperature solar-powered systems would be expected to have a negligible maintenance cost and a long life, whereas competing systems are penalised heavily by their fuel costs. Here, it is even more difficult to give a balanced comparison. It has been reckoned that small low-temperature solar plants could produce electricity at a cost to the user of about $0·05 to $0·1 per kWh. With solar pond collectors one estimate is as low as $0·02 per kWh, and some photobiological systems have been given similar figures. These costs are some two to ten times greater than the cost of producing electricity by conventional means in developed countries. However, the cost to the consumer is normally raised by a factor of at least three by the cost of transmission. In less-developed countries the cost of conventional generation is higher still, particularly in places which have no indigenous supplies of fossil fuels. Pakistan, for instance, has to import nearly all the coal used there. At the port, this may cost about $15 per metric ton, but far inland transportation costs can raise this to $100 per metric ton or more. If such a fuel is burned to produce electricity with an efficiency of 25%, the cost attributed to fuel costs alone is then about $0·03 per kWh. Similar situations occur in many parts of the world.

Detailed studies, related to particular types of installation in particular locations, are required before substance can be given to the bare skeleton of data provided here. This is enough, however, to render plausible the conclusions of many recent studies that solar-powered systems can now be considered to be fully competitive with others in several parts of the world.

Some other devices deriving their energy from the sun, but not used for power production, are in a more favourable, even an advantageous position. The solar water heater is in very wide-spread use in certain countries, particularly in Japan and Israel, which have no indigenous fuel supplies, but also in parts of the U.S.A. and U.S.S.R. There has yet been little extension of its use into the vast areas of Asia, South America and Africa where it is potentially of great value. The simplest kinds of solar water heater, with the necessary storage tanks, have a prime cost of some $15 per square metre of collector area; even in the southern U.S.A. a rather more complicated and efficient system can be provided for $50 per m² or less. The prime cost of a simple solar system capable of providing most of a household's needs would be less than $200 in most parts of the world. Nearly everywhere between the 40° latitudes it would provide hot water at less than half the cost of heating it by electricity, gas or solid fuel. The saving in fuel by such a device can generally pay off its cost in about 5 years.

A large extension can be foreseen for the use of solar driers for fish, fruit and vegetables wherever these are grown. These have already been constructed for a first cost of some $10–15 per m² of surface and have recovered their costs in 2–3 years of operation in the drying of crops such as apricots and coffee. Solar water stills cost a similar sum at present, though the development of a plastic film for covers with a satisfactory life is likely to lower this further. The cost of producing potable water by this means is about $½ per m³ at present. Although this is about ten times the price paid in developed countries with a damp climate, such as Britain, where water is cheap, it is attractive in many other places. Stills share with solar driers the advantages that they can be constructed from commonplace materials, wherever they are needed and in the sizes required. The lack of any need for collective action, central planning or the setting up of costly distribution networks is greatly in their favour.

Several studies have been made of the production of solar

cookers at minimum cost in various parts of the world. These have shown that a device suitable for a family, say providing $\frac{1}{4}$ to $\frac{1}{2}$ kW maximum power, cannot be produced for much less than about $10 at present. Although this seems small for a device with a 10-year life, efforts to encourage their use on a large scale have not been notably successful.

The developments we have considered here are part of a steadily-developing pattern. The unguarded (and largely unheeded) optimism of advocates of solar power in the past is now giving way to sober engineering and economic analysis. This is showing that there is already a place for sun-powered systems and their relative position is bound to improve further. It remains to be shown how this can best be encouraged.

10.5 *A place for sun-powered systems*

We have seen that we cannot expect solar power plants to be competitive with conventional systems for the production of power on a large scale. However, we noted in Chapter 1 that there are reasons for believing that the large-scale production of power need not be a serious problem for mankind, though it may be costly. Far more attention needs to be given to the provision of power on a small scale in scattered and isolated places. It is here that solar-power systems could have a major influence on economic development. We saw in the previous section that modest amounts of power can be furnished by low-temperature systems at a competitive price. The penalty is the high capital cost, about $1000 per kW of capacity. This is the inevitable result of the very low intensity of solar radiation and no great improvement on it is to be expected. This does not mean, however, that there is unlikely to be a market for such systems. The widespread use of small wind-driven pumps and generators, having similar characteristics to solar systems, suggests that low maintenance and absence of fuel costs are powerful factors in the attraction of such devices in farming communities. It is probable that a substantial market is to be found already in the U.S.A., U.S.S.R., Southern Europe, South Africa, Japan, Australia and New Zealand. Certainly, there are enough indications to justify serious market surveys in these areas, and there are prospects for the establishment of a manufacturing industry in this field—perhaps in southern Europe and in Australia.

There are even wider prospects for an increasing variety of the other kinds of devices using solar energy which we have discussed; the water heaters, stills, driers and so on. These could profitably be employed, not only in the countries already mentioned, but in regions of many others, such as South America, the Caribbean, Africa, India and South-East Asia. The market here, once established, should continue to expand due to the increasing efficiency of the activities concerned and would help to spread economic advancement into surrounding territory.

But in Chapter 1, it was suggested that one of the greatest tasks in human advancement is the support and stabilisation of the rural population in underdeveloped countries. Where there is the greatest need and the greatest effect for a given investment there is also the greatest difficulty. Anyone in the position of a barely-subsisting peasant cannot afford $1 000 per kW for power plant, even if it is cheaper in the long term than draught animals. The number of persons with an annual income of $100 or less may be as great as a thousand million. They cannot afford *anything*. Moreover, they cannot lightly risk the slightest departure from the way of life established by their forebears over the centuries. Even where it can clearly be shown that a course of action is economically better than existing procedures, there is no background of investment, buying out of income, loan structures—even no history of possessing material property— that western civilisation takes for granted. Patterns of land tenure, rules of succession and multitudes of quasi-legal, religious and social factors now become important and all tend to preserve the current tenor of existence. The problem ceases to be a purely technical and economic one and becomes a sociological one.

To attempt a solution would be to move outside our present terms of reference. It must clearly be part of much wider questions on the best use of resources in order to create a development pattern which is self-propagating. But at least the utilisation of solar energy offers a sensible way of using financial and technological aid which has been rather neglected and calls for serious study. It may be questioned whether it is desirable to donate devices made by overseas industry, supposing such an industry to have developed along lines suggested as plausible here. It is probably better to support the development of indigenous solar-energy industries. These might be backed by firm Government orders for initial free or subsidised issue of plant for use in remote areas. Some of this might be in kit form, calling for a limited amount of local effort in assembly. Legislation requiring the use of solar energy-converting devices, water heaters, coolers and so on in hospitals, schools, factories, Government buildings or even in all buildings in more developed areas, would make the first step towards the establishment of a self-supporting industry. In turn, through increased exports and lowering of imports of fuel, a general economic improvement could follow.

These thoughts will remain mere speculation without extensive further study of the possibilities in particular areas. But the present work can provide a basis for the technical appraisal of simple solar energy-converters, which must be an essential part of all such studies. Through these, another step might be taken on the long road towards equality of opportunity for everyone, wherever we happen to live. The sun shines without partiality upon us all.

SUGGESTIONS FOR FURTHER READING

For an extension of the treatment of some of the topics in this book, the reader might begin with

Zarem, A. M. and Erway, D. D. *Introduction to the utilization of solar energy*. McGraw-Hill, 1963.

which also provides a good bibliography.

Articles and research papers are to be found everywhere in the literature of science and engineering. In recent years, one of the main sources has been the journal

Solar Energy, published for the Solar Energy Society, Arizona State University, Tempe, Arizona.

European writers contribute also to the journal of the

Coopération Mediterranéenne pour l'Energie Solaire, (COMPLES) published under the auspices of the Faculty of Science of the University of Aix-Marseille.

The proceedings of international conferences bring together much of the original work from all parts of the world. Two conferences organised by the United Nations Organisation have related to solar energy and its application:

UNESCO Symposium on Solar Energy and Wind Power in the Arid Zones, New Delhi, India, 1954.

U.N. Conference on New Sources of Energy, Rome, 1961.

More papers are to be found in the transactions of the NATO symposium

Solar and Aeolian Energy, Sounion, Greece, 1961.

Two conferences organised in 1955 provide a further range of studies in utilisation of solar energy:

Conference on Applied Solar Energy, Tucson, Arizona (Transactions by the University of Arizona Press, 1958)

World Symposium on Applied Solar Energy, Phoenix, Arizona (Proceedings by the Stanford Research Institute, 1956).

At the time of going to press, the Solar Energy Society is arranging an international conference in Melbourne, Australia in March, 1970.

NAME INDEX

SUBJECT INDEX

absorption, 29, 68
absorptivity, 68, 72
activity, 217
adiathermal, 117, 128
air mass, 36
algae, 225
altitude, solar, 21
anode, 150, 160
azimuth, solar, 21

bacteria, 226, 228
bilharziasis, 7
black body, 69
Boltzmann constant, 57
bonding, 216

Carnot efficiency, 124, 170, 179
Carnot cycle, 128
cathode, 150, 159
chlorophyll, 221
cloud cover, 29, 232
coal, 9, 14
coefficient of performance, 140
concentration ratio, 79, 83
coulomb, 149
Coulomb force, 65, 149, 216
current density, 163
cut-off, 188
cycle, 118

deactivation, 211
declination, solar, 32
depletion layer, 192
desalination, 13
deserts, 43
diathermanous, 68
diffuse radiation, 38
direct radiation, 38
distillation, 105
doubling distance, 21
doping, 188
duality, 47

efficiency, 121
electrochemical cell, 215, 218
electrode potential, 218
electrolyte, 215
electrolysis, 207, 238
electron spin, 212
electron volt, 158
emissivity, 72
energy band, 153
energy demand, 10, 11, 13
energy density, 26, 70
energy, internal, 52
energy, kinetic, 51
energy level, 66
energy, potential, 51
energy spectrum, 26, 187
enzyme, 220
equation of state, 58
equilibrium, thermal, 54
equinox, 21
Ericsson cycle, 131
exclusion principle, 153

Fermi-Dirac distribution, 158
Fermi level, 158
fertilisers, 13
figure of merit, 178
filariasis, 7
first law of thermodynamics, 118
fission, 15
flat plate collector, 73, 92
fusion, 15, 24

gamma radiation, 25

heat, 52
heat pump, 121
heliostat, 90
hole, 190
hour angle, solar, 32
humidity, 108
hydroelectricity, 17

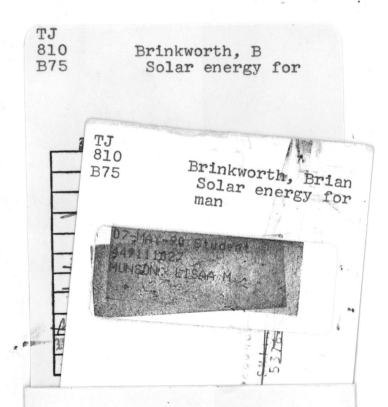